The Art of Theoretical Biology

Franziska Matthäus • Sebastian Matthäus • Sarah Harris • Thomas Hillen
Editors

The Art of Theoretical Biology

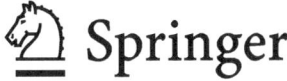 Springer

Editors
Franziska Matthäus
Center for Computational and Theoretical Biology
University of Würzburg
Würzburg, Germany

Sarah Harris
Physics and Astronomy
University of Leeds
Leeds, UK

Sebastian Matthäus
Grenfarben Agentur für Gestaltung
Berlin, Germany

Thomas Hillen
Department of Mathematical and Statistical Sciences
University of Alberta
Edmonton, AB, Canada

ISBN 978-3-030-33470-3   ISBN 978-3-030-33471-0   (eBook)
https://doi.org/10.1007/978-3-030-33471-0

This Springer imprint is published by the registered company Springer Nature Switzerland AG
The registered company address is: Gewerbestrasse 11, 6330 Cham, Switzerland

The Art of Theoretical Biology

Preface

Every image in this book was created for a purely scientific purpose. The images were obtained using mathematical modelling and simulations to explore and understand biological or medical concepts. The images were chosen since they touched the scientists on a different level. They inspire further thought, they showcase disturbing facts, express confusion, or are simply aesthetically beautiful. By being included in this collection, they have won the honourable title of "art".

Some think this accolade must be reserved for works created out of the sheer joy of creativity alone. These images, however, all share a fundamental purpose, they are based on research in theoretical biology.

Each contributing author has provided a personal explanation of the underlying science they were exploring. Many of them describe why their chosen image was so striking to them. This indication of purpose adds resonance to the pictures, rather than taking away their mystery. The images show how beauty can arise in the darkest of places, such as in the breakdown of cell function during cancer, and how complexity generates richness in unexpected ways. Each picture provided their creators with an inspiration distinct from the purely scientific results they originally intended. Therefore, we view the book as a superposition of art and science; each separate image is an act of scientific research, but the whole collection is a work of art.

The field of Theoretical Biology has made some fundamental contributions to our understanding of nature, biology, and medicine. Within our community we study the impact of global warming on polar bears, reindeer populations, migrating birds, and mountain pine beetles. We study the spread of diseases and advise governments about vaccination strategies, we develop pharmaceutical drugs, medical procedures and optimize cancer treatments, we visualize brain tumours, bones and organs, and we help to understand how individual cells move, how they polarize and divide, how they react to signals from their environment and how it is possible for individual cells to produce a living, conscious species such as humans. Theoretical Biologists use advanced tools from mathematics and computational sciences, which are targeted to the biological problem at hand.

Several of the contributions relate to active research in Oncology. Cancer is a devastating illness and many of us have had first hand experience with this disease. The cancer related contributions of this special issue are chosen to show our deep respect to all people that are affected by cancer. The contributions showcase the combined efforts of the research community to better understand the effects of cancer and to design more effective treatments. Each of the contributors is driven by the goal to make a substantial impact on cancer treatment, to reduce cancer related death, and to improve the quality of patients' lives. The contributions also show that cancer researchers are simply humans, people who deeply care and who appreciate some form of beauty in an otherwise disturbing research topic.

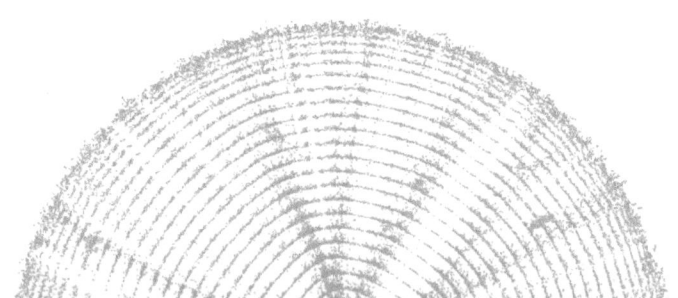

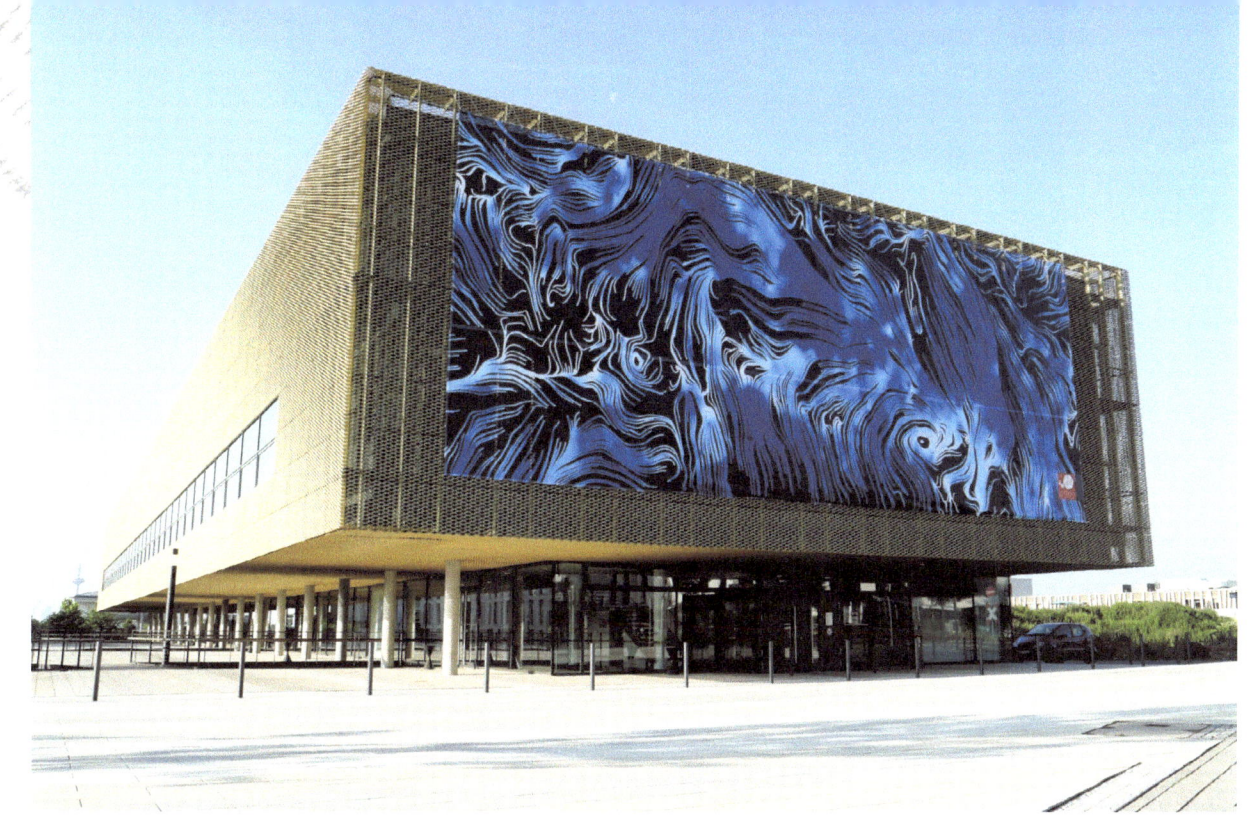

This book is comprised of 71 contributions, involving more than 120 authors. The big adventure of bringing the book into existence would not have been possible without the enthusiasm, support and, not the least, the patience of all of our authors, for which we would like to thank them deeply. Since the first discussion of the editorial team in the early spring of 2016 several years passed, during which we issued repeated calls for contributions, went through selection rounds and collected abstracts. We also address our deepest thanks to Jan-Philip Schmidt (SpringerNature Publishing), who supported the project during the entire period. And last, but not least, having the collection and a publishing house, we still faced the challenge of bringing all these different images into one coherent product. This involved formatting of all images for book print, design aspects of how to present the images, as well as their variety, in an optimal way, and the creation of a layout for the text parts and the book cover. Fortunately, Sebastian Matthäus, head of a graphic design agency located in Berlin, Germany, agreed to support us with his expertise.

With his help we organized an exhibition, starring a subset of our images, at the Riedberg Campus of the University of Frankfurt in June and July, 2018. This exhibition was only possible due to generous financial support from the Frankfurt Institute for Advanced Studies, allowing us to display large-scale printouts on the occasion of the Frankfurt Night of Science. Further exhibitions will follow.

For each of our images you can find a description of the scientific context and some scientific publications are listed. The author's affiliations are provided as well at the back of this book. Feel free to contact any of them if you would like to learn more about a specific topic, or if you would like to offer your support. Our authors are happy to respond.

Enjoy browsing through these images. Stop at anything that touches you and speculate what it could mean. Then read the description. Enjoy.

Franziska Matthäus, Sebastian Matthäus,
Sarah Harris & Thomas Hillen

Content

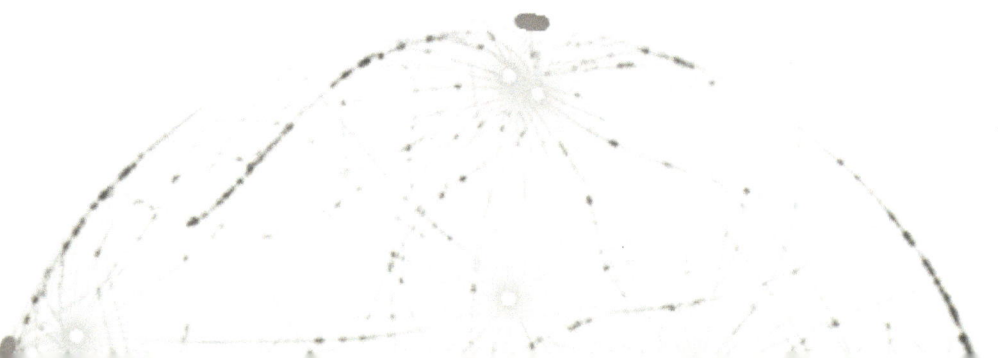

Content

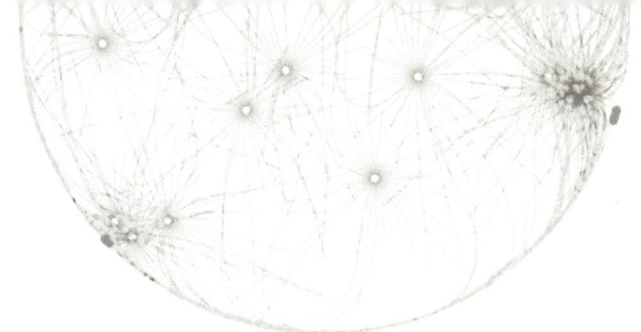

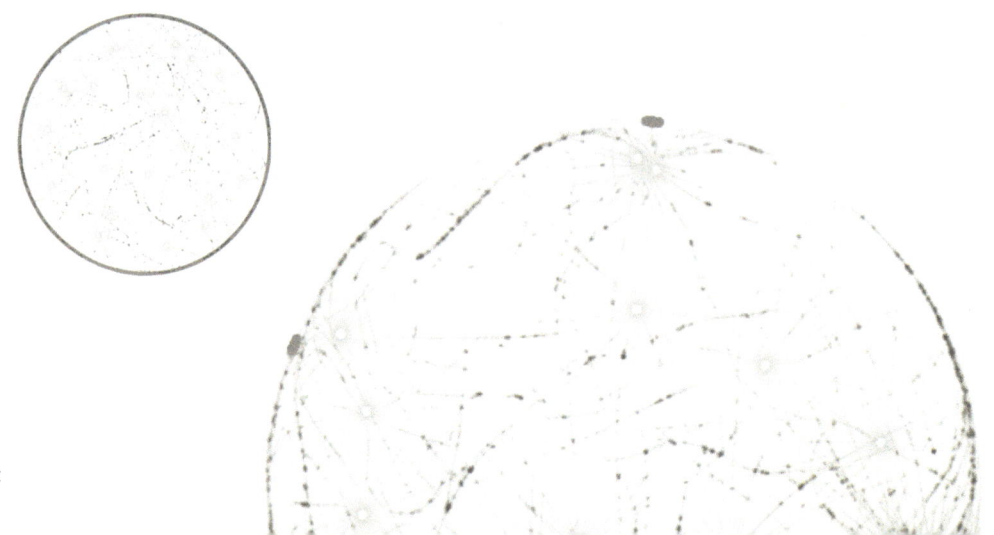

It makes you wonder ...

The Deadly Beauty of Cancer

By Bartlomiej Waclaw & Martin Nowak

The research story

Cancerous tumours are not a uniform mass of identical cells but can be very heterogeneous [1]. Cells from different parts of the tumour harbour different genetic mutations. These mutations affect how fast cancer cells are able to grow, whether they can invade the surrounding healthy tissue, or how sensitive they are to chemotherapy. Heterogeneity not only makes tumours more difficult to eradicate but also to diagnose, because a small biopsy sample may not be representative of the entire tumour. We were interested in what biological processes determine how heterogeneous tumours are. We developed a computer model that simulated a population of cancer cells that replicate, die, migrate, and mutate. We varied the strength of these processes and measured the level of heterogeneity.

The image

The image shows a simulated tumour with a low level of migration. Cells are represented by small dots. The tumour shown in the foreground has about 10 million cells, smaller tumours in the background are snapshots of the same tumour from earlier times. Cells have been colour-coded depending on what mutations they carry in addition to the first cancer-initiating mutation. Cells with similar mutations have been assigned similar colours. A huge diversity of colours means that the tumour is genetically heterogeneous. This is typical for our simulation. Only when migration is very fast or cells rapidly die and are replaced by other cells, tumours become more homogeneous. The sequence of images has been created using a published computer algorithm [2].

References

[1] Gerlinger M et al., Intratumor heterogeneity and branched evolution revealed by multiregion sequencing, New England Journal of Medicine 366: 883–892, 2012.

[2] Waclaw B et al., A spatial model predicts that dispersal and cell turnover limit intratumour heterogeneity, Nature 525: 261–264, 2015.

© Springer Nature Switzerland AG 2020
F. Matthäus et al. (eds.), *The Art of Theoretical Biology*, https://doi.org/10.1007/978-3-030-33471-0_1

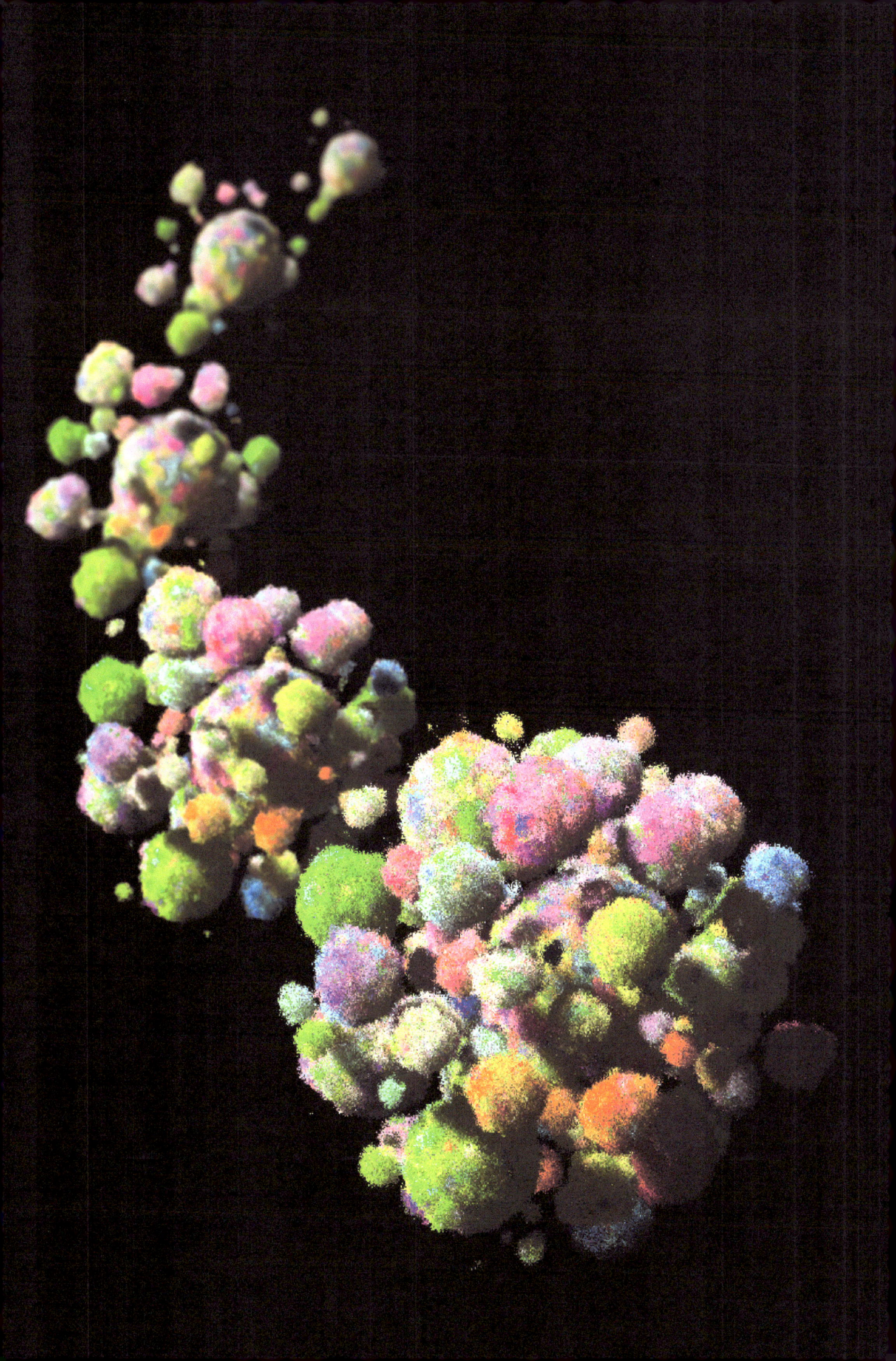

Cellular Connections

By Roeland Merks

The research story

"Cellular Connections" shows a computer simulation of the growth of blood vessels. Endothelial cells, the building blocks of blood vessels, collectively form networks of blood vessels, much like ants work together to form their nests. In the centre of the image, the cells are sufficiently close together such that they feel one another and manage to interconnect. The cells at the periphery are too far away and wander around aimlessly.

The image

The tiny blood vessels that we simulate are formed early in embryonic development, and they continue to grow throughout our lives. Cells can stimulate adjacent blood vessels to form side branches as a healthy response to lack of oxygen, for example during wound healing and menstruation. Unfortunately, cancer cells can hijack this process, and attract blood vessels for their own benefit. By figuring out the rules that endothelial cells use to construct a blood vessel, we hope to find new ways to control blood vessel growth. Through this computer simulation we found out that the cells form particularly realistic networks if we let them assume an elongated shape, and we let them attract one another through a signal that they emit into the environment. The simulations are performed on a square grid through stereotypic algorithmic steps. Yet they generate life-like forms. This is due to the biological model rules and a pinch of mathematically-generated unpredictability.

References

[1] Merks RMH, Brodsky SV, Goligorksy MS, Newman SA, Glazier JA, Cell elongation is key to in silico replication of in vitro vasculogenesis and subsequent remodeling, Dev. Biol. 289(1): 44–54, 2006.

[2] Palm MM, Merks RMH, Vascular networks due to dynamically arrested crystalline ordering of elongated cells, Phys. Rev. E 87: 012725, 2013.

© Springer Nature Switzerland AG 2020
F. Matthäus et al. (eds.), *The Art of Theoretical Biology*, https://doi.org/10.1007/978-3-030-33471-0_2

Annealing Party

By Maximilian Strobl & Daniel Barker

The research story

A phylogeny shows the pattern of relationships for a set of species – their family tree. Traditionally used in taxonomy and evolutionary biology, the algorithms used to reconstruct phylogenies from DNA sequencing data are now also finding applications in cancer research and epidemiology. However, inferring a phylogeny is a challenging optimisation problem: Out of the many possible ways in which the species might be evolutionarily related, one must identify the most plausible one. This is analogous to trying to climb to the top of the highest peak in a large mountain range, in fog, without access to a map of the area. In this project, we studied how the simulated annealing algorithm tackles this problem, to learn more about the algorithm and the topography of the search landscapes.

The image

Shown here is an attempt to map the path of the algorithm for one data set. Each point represents one candidate phylogeny considered by the algorithm, and its colour indicates if it was considered early (red) or late in the search (blue). Distance reflects the similarity between two phylogenies. The data was collected by running a modified version of the LVB phylogeny reconstruction software, and visualised using the treesetviz package for Mesquite [1]. One can nicely see how the search starts by exploring disconnected regions, but then converges to a final plateau. Redrawn from Strobl and Barker (2016) [2].

References

[1] Hillis DM, Heath TA, St. John K, Analysis and visualization of tree spaces, Systematic Biology 54: 471–481, 2005.

[2] Strobl MAR, Barker D, On simulated annealing phase transitions in phylogeny reconstruction, Molecular Phylogenetics and Evolution 101: 46–55, 2016.

© Springer Nature Switzerland AG 2020
F. Matthäus et al. (eds.), *The Art of Theoretical Biology*, https://doi.org/10.1007/978-3-030-33471-0_3

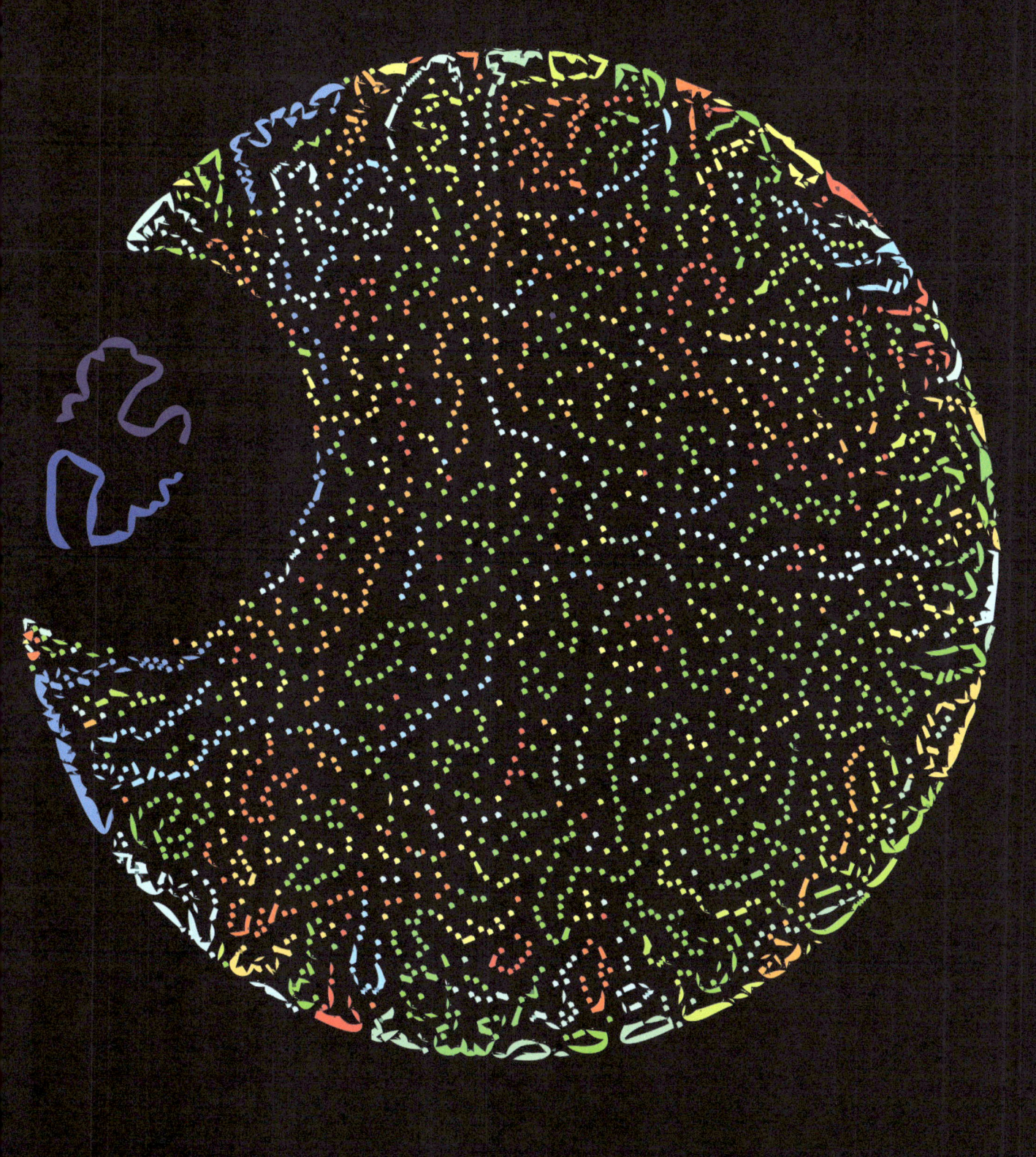

Cells on the Ferris Wheel

By Katharina Becker

The research story

When objects flock together and keep moving they often start to rotate. From penguins huddling together to keep warm to cells that move towards a chemical attractant this phenomenon is wide spread in nature. While penguins are arguably more entertaining to observe, our current research focuses on cellular aggregation in developmental processes. During the formation of hair or feather follicles, cells in the skin compact. Their movement is determined by the availability of growth factors, that are secreted by the cells themselves and act as chemoattractants. The cells actively migrate towards higher concentrations of these substances and aggregate. The dense packing of cells gives rise to mechanical forces where compressed cells repel each other and thus limit local chemotactic effects. We use an agent-based model to simulate these interactions.

The image

The image shows results of these simulations for different combinations of mechanical and chemotactic effects. The interplay of repulsion, cell-cell adhesion and chemotaxis determines whether clusters rotate. Each orange circle represents a cell. Blue arrows indicate their velocities while red arrows show the sum of the forces acting on the respective cell. The panels on the left show cells whose active migration is determined by both the chemoattractant and the forces acting on them. Their velocity vectors point directly towards the centre of the cluster; they do not rotate. The panels in the middle show adhering cells, those on the right non-adhering cells whose active migration is mainly determined through chemotaxis. Their velocity vectors are more tangential indicating cluster rotation.

© Springer Nature Switzerland AG 2020
F. Matthäus et al. (eds.), *The Art of Theoretical Biology*, https://doi.org/10.1007/978-3-030-33471-0_4

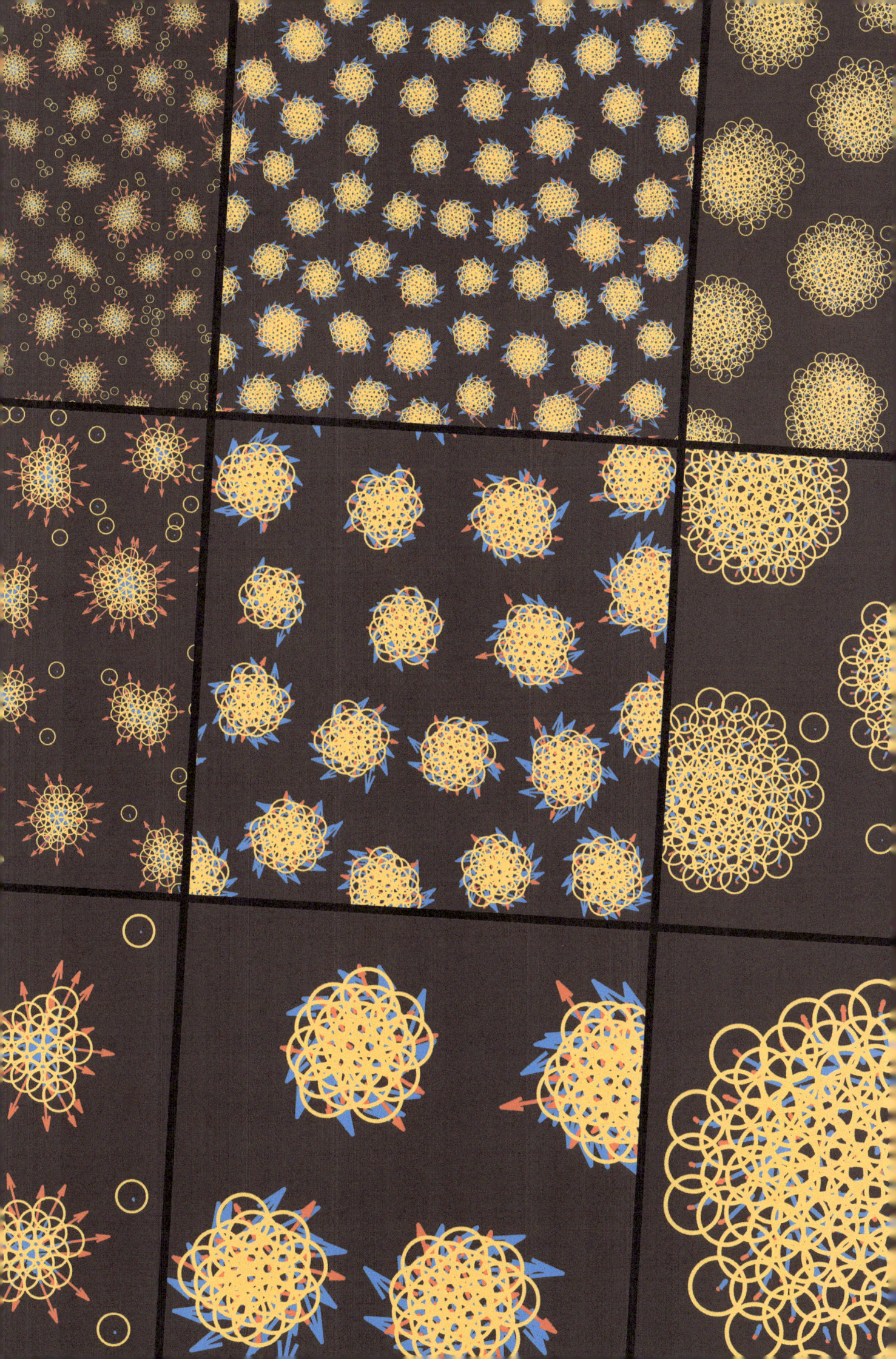

The Magic Pants that Always Fit

By François Nédélec

The research story

The inner surface of animal cells is covered by a filamentous scaffold called the cortex, which is composed of micro filaments connected by proteins such as molecular motors. The filaments are made by assembly of a protein called actin, and typically can reach a length of a few micrometers. The connecting proteins are typically much smaller, but some molecular motors can be up to hundreds of nano-meters in size.

The image

This picture shows a cortex covering the entire cell, where the filaments are connected to each other by invisible molecules. It is a numerical model made with Cytosim, an Open Source software constructed to simulate cellular structures such as this one (www.cytosim.org). The cell was here subjected to a virtual "osmotic shock" in the form of an overall volume decrease. Wrinkles formed as the shell was compressed, as depicted on the image. In a real living cell, however, such wrinkles would rapidly vanish because the filaments naturally disappear while new ones are created. Like many other structures in the cell, the cortex is always dynamic and can thus readily readjust following a change of volume. Thus it fits the cell always perfectly! Many other cellular structures are similarly dynamic allowing them to adapt to conditions that are often different between various cells in the body.

© Springer Nature Switzerland AG 2020
F. Matthäus et al. (eds.), *The Art of Theoretical Biology*, https://doi.org/10.1007/978-3-030-33471-0_5

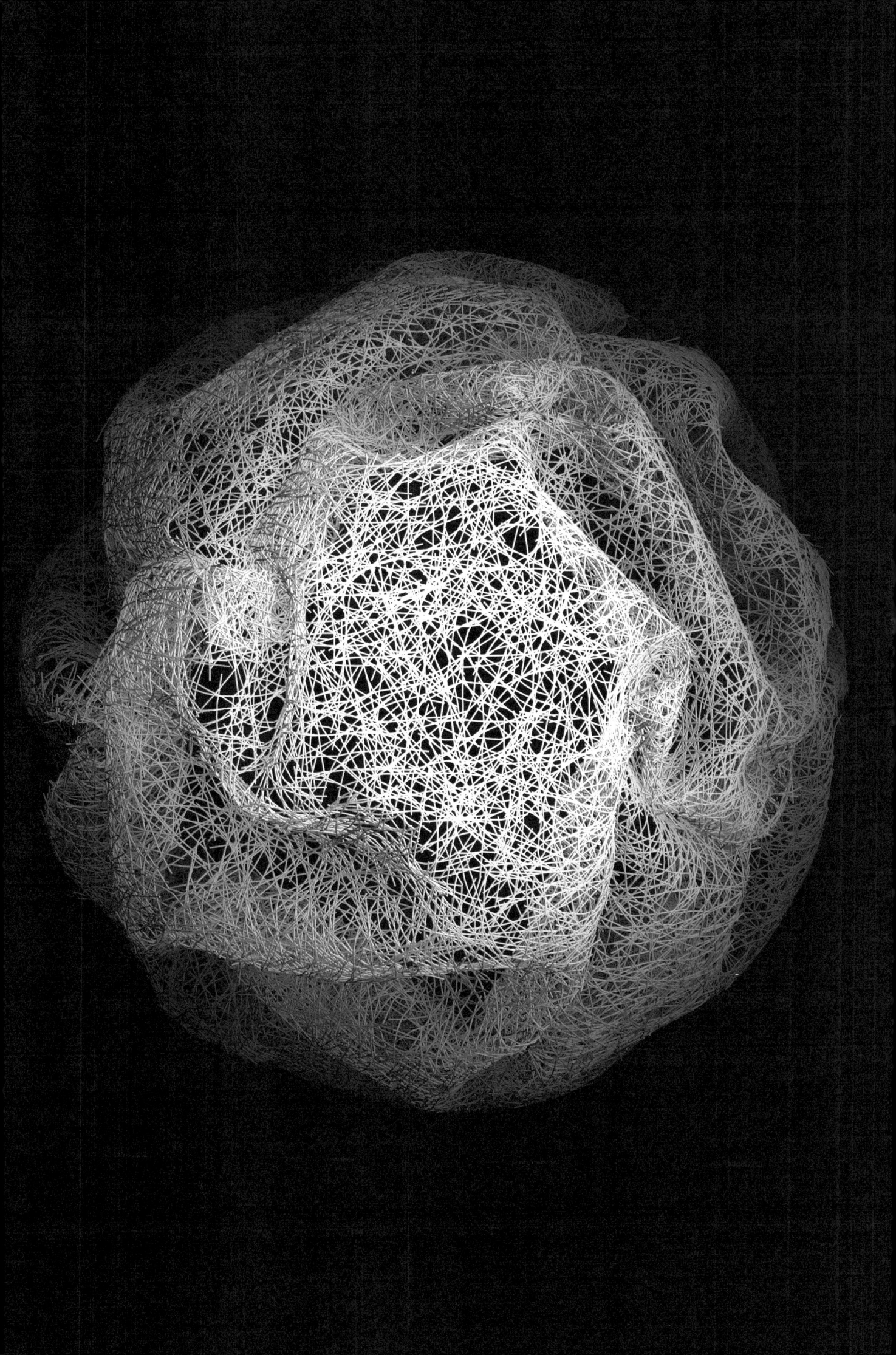

Racing Triangles

By Glenn Webb

The research story

One way to mathematically model population growth structured by the age of individuals is to design discrete age classes, which evolve in discrete time steps. This model uses a transition device, known as a Leslie matrix [1], to provide a description of how each age class changes at each time level. At each time step, the population is represented according to the number of individuals in each age class at that time. The Leslie matrix successively generates the population age distribution at each next time level. The Leslie matrix incorporates information about the fraction of individuals that survives from one age class to the next age class, and the number of offspring produced by an individual in each age class. The behavior of the population over time is determined by properties of the Leslie matrix known as eigenvalues. If all of the eigenvalues are less than 1, then the population decreases. If at least one of the eigenvalues is greater than 1, then the population will grow.

The image

The figure represents an age-structured population with three age classes. One eigenvalue of the corresponding Leslie matrix is greater than 1. The population exhibits increasing growth with period-three oscillations. The triangles represent the period three oscillations, with the length of each side of a triangle corresponding to one of the three age classes. The increasing size of each triangle corresponds to the overall growth of the total population as time evolves.

Reference

[1]   Leslie PH, The use of matrices in certain population mathematics, Biometrika 33: 183–212, 1945.

F. Matthäus et al. (eds.), *The Art of Theoretical Biology*, https://doi.org/10.1007/978-3-030-33471-0_6

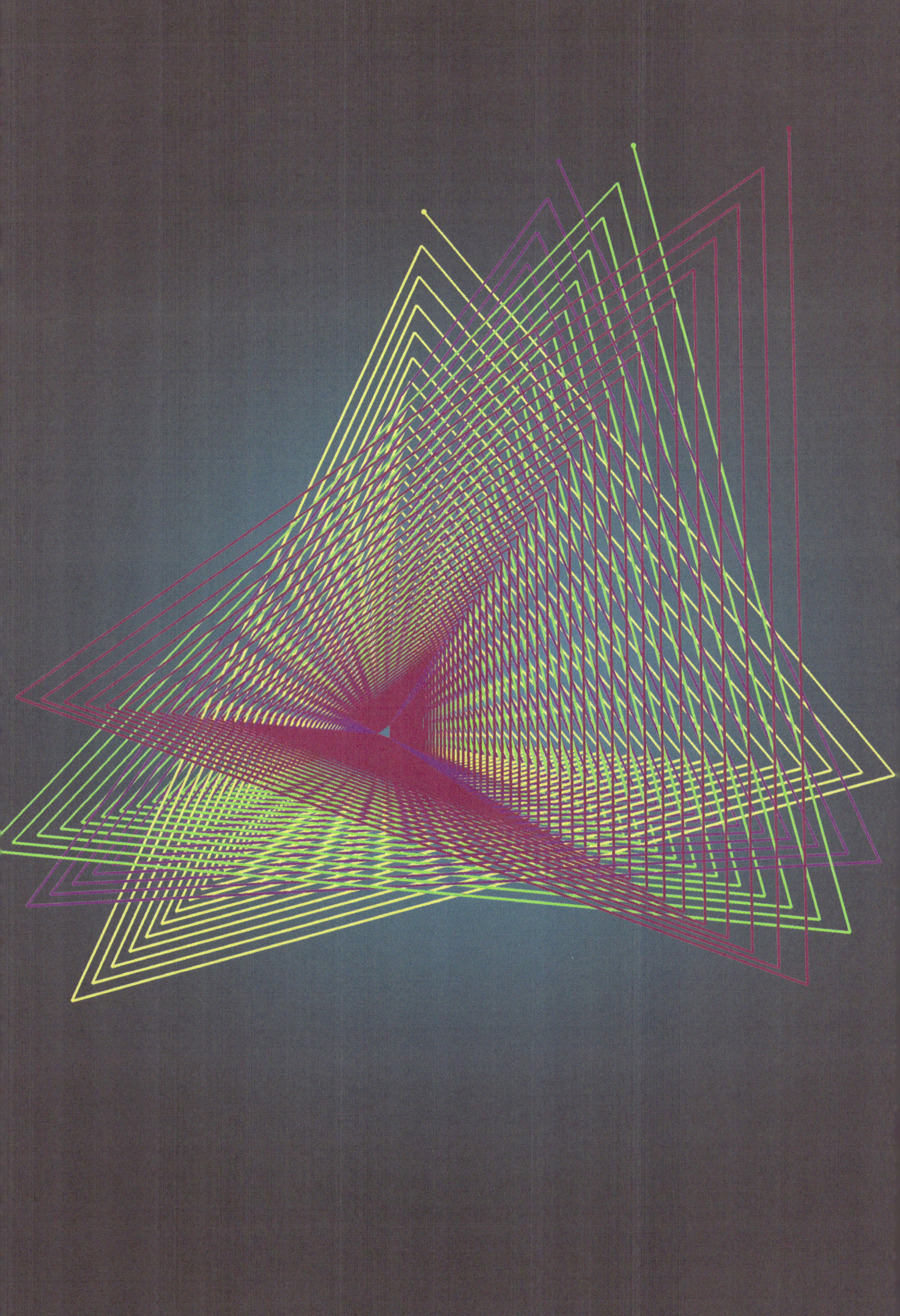

Rising Dragons

By Vikram Adhikarla, Daniel Abler, Davide Maestrini,
Russell Rockne & Prativa Sahoo

The research story

There were two pretty dragons Hydra and Medusa who were known in ancient times as the best friends around. Once they had a small tiff with each claiming to be more beautiful and colorful. The left one is Hydra and the right one is Medusa. Can you tell who's more beautiful? So how did Hydra and Medusa come into being? Yes! That's correct! From stem cells moving stochastically in the brain and getting directional cues from tissue structure. A histological section of a mouse brain stained with a white matter recognizing agent was imaged and Structure Tensor Analysis (STA) was performed on it. STA generates the dominant eigenvector and coherency of a particular tissue pixel indicating the directionality and the degree of anisotropy of that pixel. White matter tracts have directional information and higher coherency. If seeds are initialized in the white matter (corpus callosum) then they move along the dominant eigenvector. The goal is to evaluate the migration characteristics of stem cells in the brain to optimize cellular therapies.

The image

The left half of the image is a mirror image of the right half to make it look cool. So both Hydra and Medusa were actually clones! The individual tracks of the initialized seeds are shown in different colors. The region with a high density of paths is where the seeds were initialized. STA was performed in ImageJ and the ellipses (orange) and the dominant eigenvector directions (lines) are shown.

© Springer Nature Switzerland AG 2020
F. Matthäus et al. (eds.), *The Art of Theoretical Biology*, https://doi.org/10.1007/978-3-030-33471-0_7

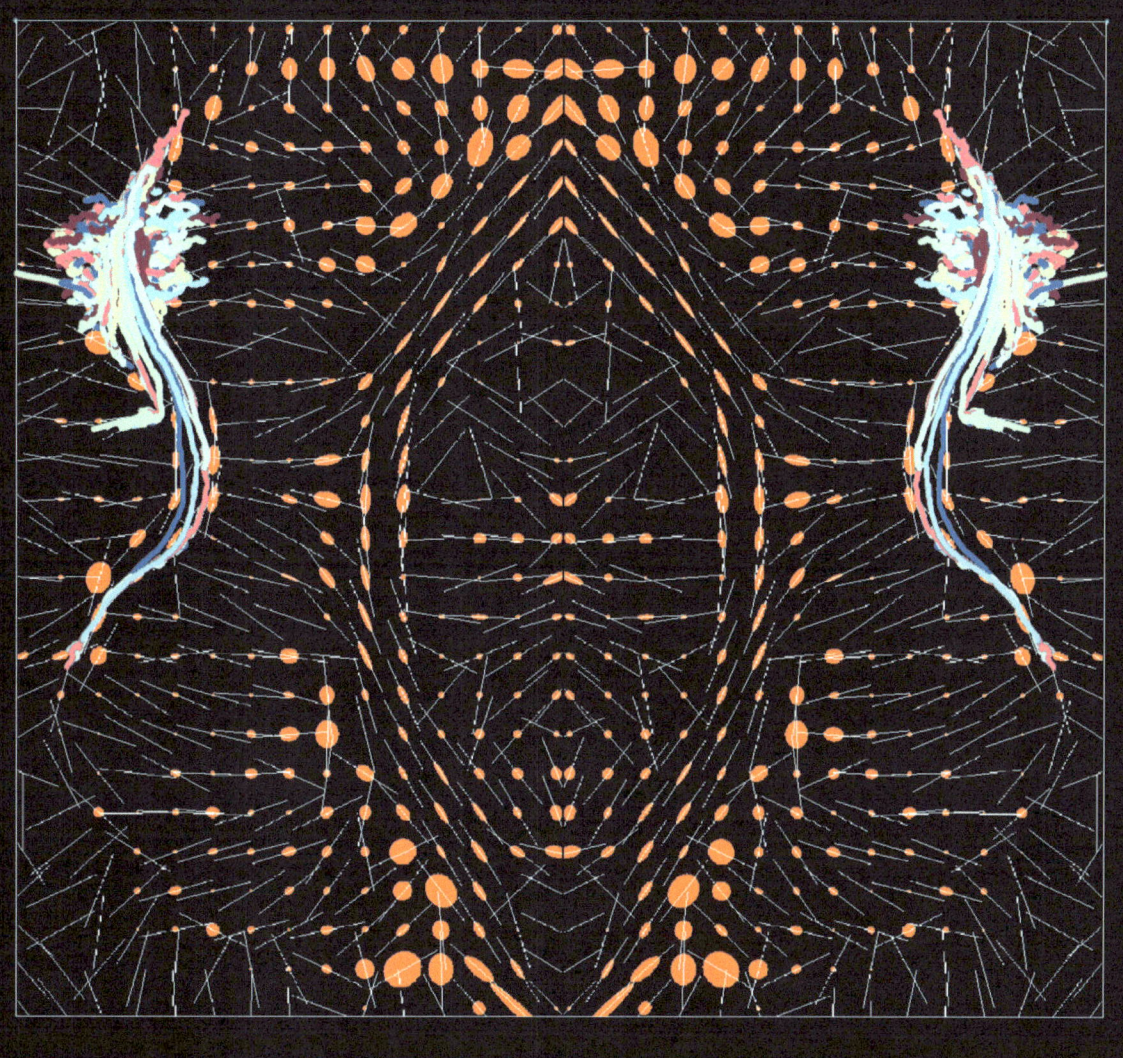

Henri in Wonderland

By Hanna Schenk, Chaitanya Gokhale & Arne Traulsen

The research story

We are constantly fending off parasites. As we become resistant to one parasite, we can fall ill due to a second one. As we become resistant to the second parasite, we are susceptible to the first one – back where we started. This never ending race is illustrated by the metaphor of the Red Queen (... of the Wonderland, 'Through the looking glass') explaining to Alice "it takes all the running you can do to keep in the same place". While this is true for two parasites and two types of responses, as soon as the number of parasites and the reactions to them increase in number ... chaos arises. We study three types of parasites and three types of hosts. Each parasite can successfully infect only one of the hosts.

The image

The relative abundances of hosts and parasites change over time as the infection dynamics progress. These changes are captured by six differential equations, one for each type of host and parasite. We choose a lower-dimensional section to display the dynamics: When the dynamics return to the same state as before, à la Red Queen dynamics, then it forms almost continuous lines (yellow). If the points scatter far, irregular or even chaotic dynamics (violet, blue and green) is expected. We inspect 25 different initial conditions (each denoted by a different colour). Jules Henri Poincaré (1854-1912), a French polymath, developed this technique of analysing dynamical systems, now known as the Poincaré map.

Reference

[1]   Schenk H, Traulsen A, Gokhale CS, Chaotic provinces in the kingdom of the Red Queen, J. Theo. Biol. 431: 1–10, 2017.

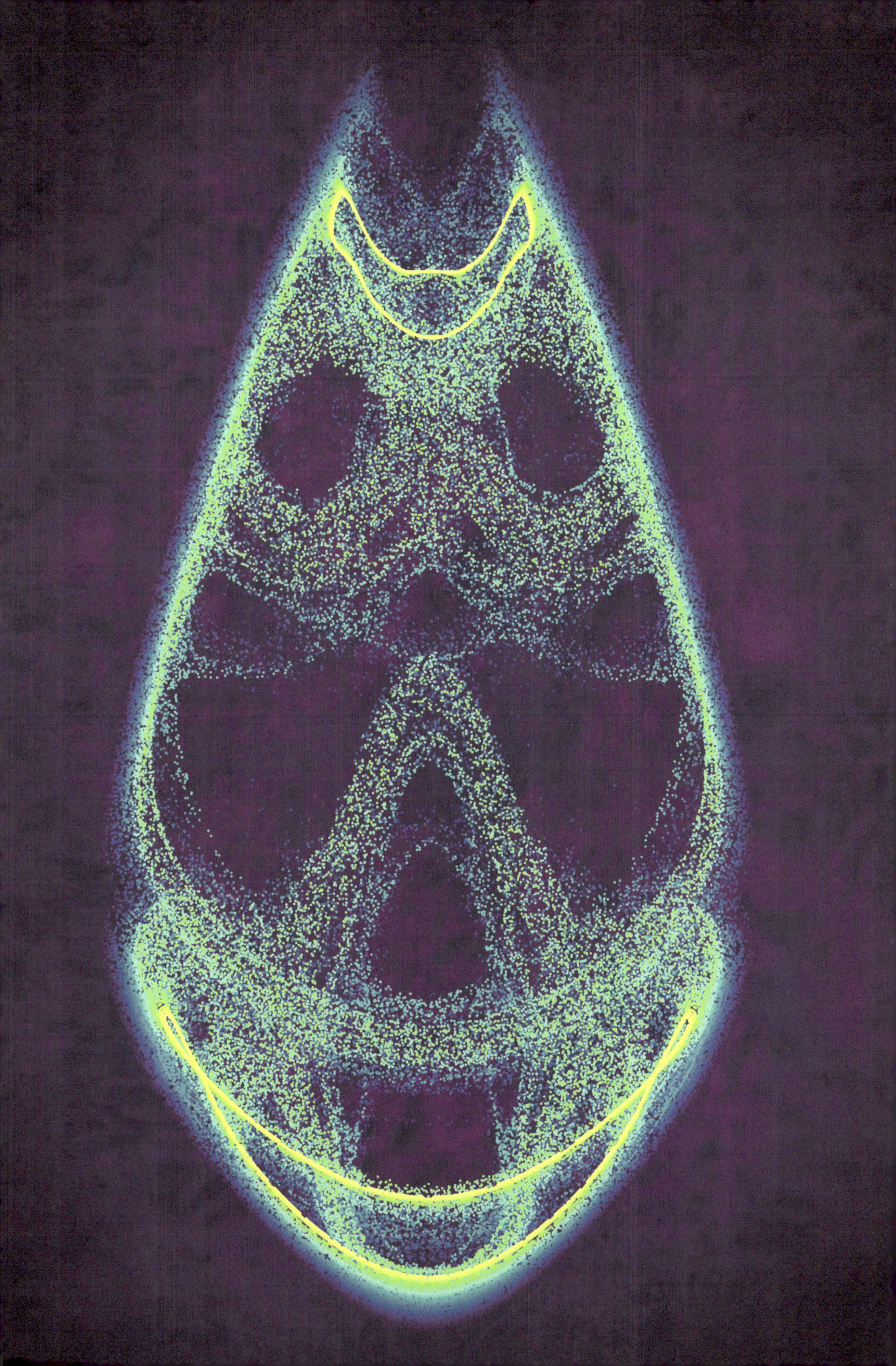

Peak of the Iceberg

By Jana Lipková, Diego Rossinelli, John Lowengrub,
Petros Koumoutsakos & Bjoern Menze

The research story

Medical imaging plays a central role in cancer therapy, however scans cannot detect the full extent of infiltrative brain tumors. Post-mortem and histological studies show that tumor cells can be found even 2 cm beyond the tumor outlines visible on the medical scans. Current radiotherapy planning is handling these uncertainties in a rather rudimentary fashion. The irradiated volume is constructed by extending the tumor regions visible on the medical scans by a uniform margin, neglecting the patient-specific tumor dynamics and brain anatomies. We calibrate a computational tumor growth model using patient-specific structural and metabolic medical scans in a Bayesian inference framework to predict tumor cell infiltrations beyond those visible on the medical images [1]. The model predictions enable personalised radiotherapy designs with improved delineation of tumor regions and identified radio-resistant areas. In turn the ensuring radiotherapy spares healthy tissue and reduces radiation toxicity, while reaching comparable accuracy with standard radiotherapy protocols [1].

The image

The visualisation shows the outline of a patient tumor visible on the medical scans (orange) together with the outline of the predicted tumour cell infiltration (blue), inside the brain anatomy (white) reconstructed from the patient's medical scans. The visualization is performed using Volume Perception [2], a volume rendering software employing a ray-casting technique. The image components are visualized as translucent isosurfaces obtained by pre-integrated volume rendering. The 3D visualizations enhance our understanding of complex tumor structures, enable the detection of possible tumor cell migration pathways, and assist in clinical decision making.

References

[1] Lipkova J, et al., Personalized radiotherapy design for glioblastoma: integrating mathematical tumor models, multimodal scans and Bayesian inference, IEEE transactions on medical imaging, 2019.

[2] Rossinelli D, Multiresolution flow simulations on multi/many-core architectures, PhD thesis, ETH Zurich, 2011.

© Springer Nature Switzerland AG 2020
F. Matthäus et al. (eds.), *The Art of Theoretical Biology*, https://doi.org/10.1007/978-3-030-33471-0_9

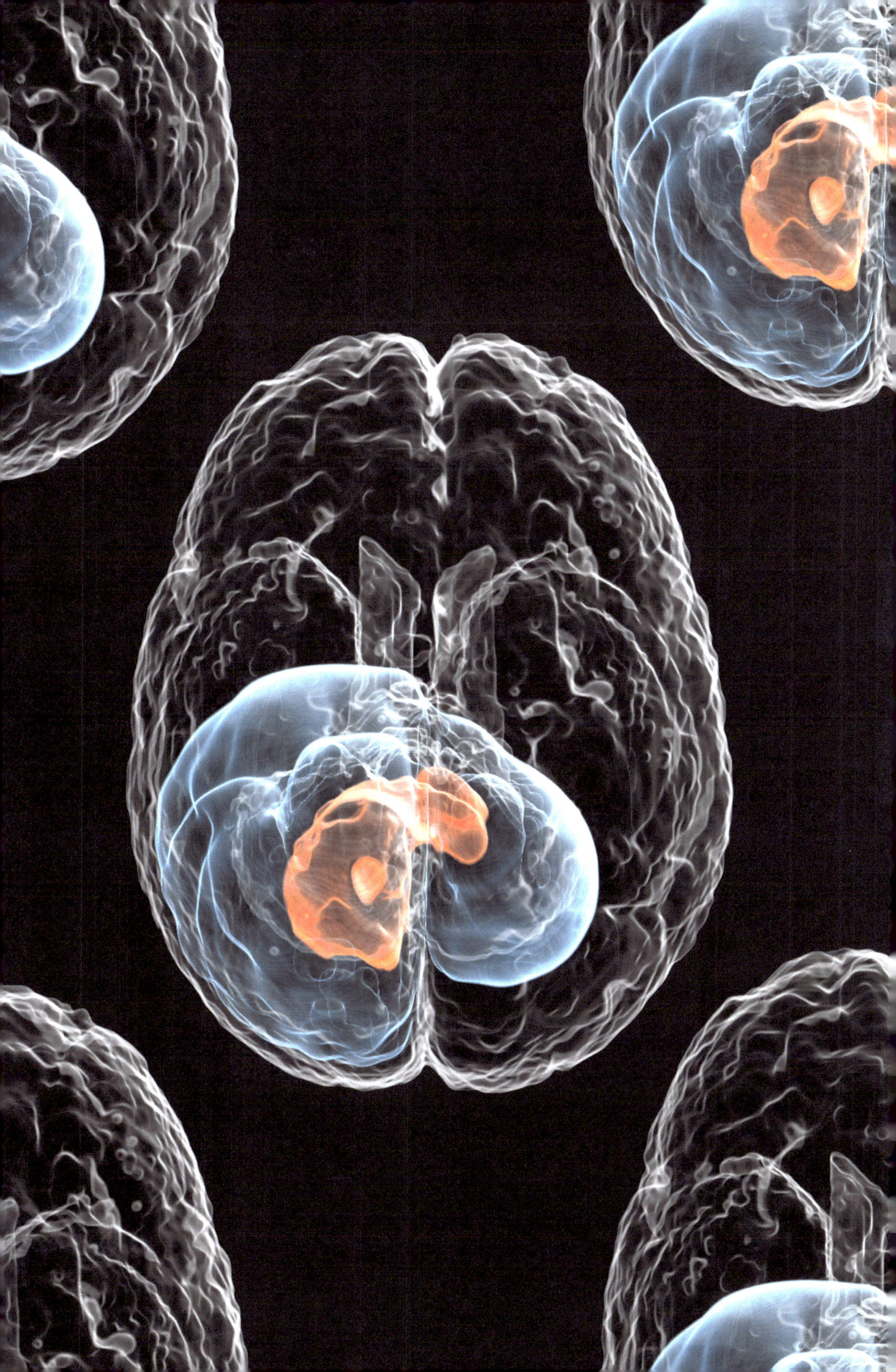

Guiding Spiral

By John Dallon & Hans Othmer

The research story

This image was generated when we studied how complex spatial patterns can result from the cell motion of Dictyostelium discoideum or Dd. Dd are amoeboid cells which when starved aggregate to form a mass of cells. In the process of aggregating, they typically form beautiful streaming patterns as they move up chemical gradients of cAMP, the chemical which guides the cells. We developed a mathematical model of this process using a system of coupled differential equations [1].

The image

This image is a visualization of a computer simulation of the solution to the equations. It shows the cell density in space with the guiding wave of cAMP superimposed. The distinctly white spiral wave is the region where the concentration of cAMP is above a threshold level. Regions where cell densities are high are shown as level sets in a rainbow-color scheme. Regions void of cells are shown in dark burgundy. A key feature required for the system to create these patterns is that the cells produce (or alter the concentration of) cAMP. This is why the spiral is rough – regions devoid of cells are not producing the chemical. The two typical patterns observed are spiral waves or concentric circles. Here cells are migrating towards the center of a spiral wave of cAMP.

Reference

[1] Dallon JC, Othmer HG, A discrete cell model with adaptive signalling for aggregation of Dictyostelium discoideum, Phil. Trans. R. Soc. Lond. B 352, 391–417, 1997.

© Springer Nature Switzerland AG 2020
F. Matthäus et al. (eds.), *The Art of Theoretical Biology*, https://doi.org/10.1007/978-3-030-33471-0_10

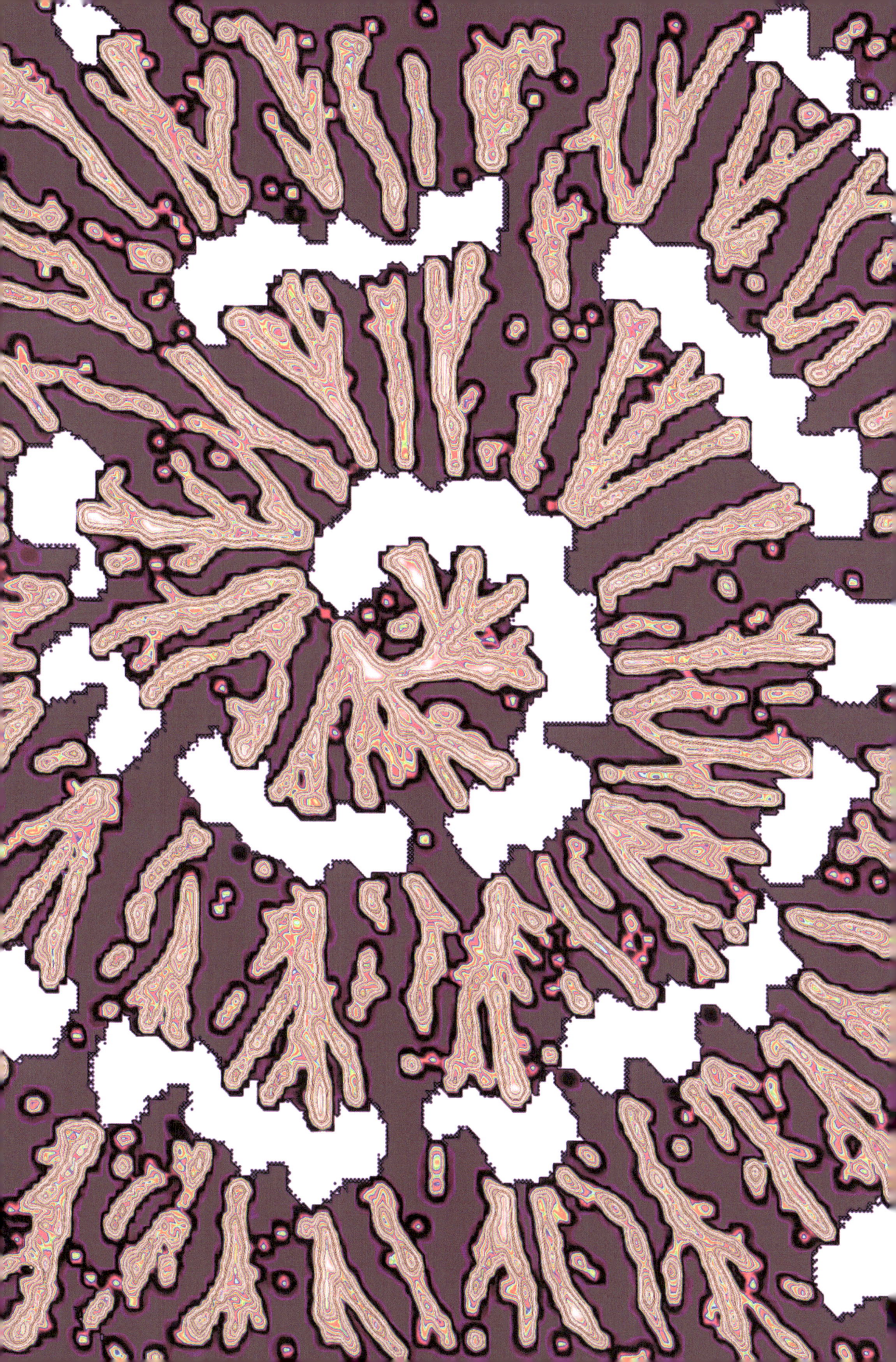

The Hidden Beauty of Roots

By Ishan Ajmera, Leah Band & Jonathan Lynch

The research story

Rice is the staple food for more than half the world's population. Increasingly limited availability of agricultural resources poses the biggest threat to rice production. Selection of rice varieties with the ability to grow in poorer soils with less water and nutrient input is therefore important for global food security. Roots are the functional unit of a plant for the water and nutrients acquisition. We are investigating how different rice plants lay down their roots and the impact this has on the plant's ability to grow with limited nutrients. From field and glasshouse experiments, the number, length, angle and growth rates of different types of roots and root structures are recorded. This data is fed into the computer-based model to predict the best type of root structures with maximum nutrient uptake and optimum plant growth in nutrient limiting conditions. The findings from this model will benefit the rice breeding programmes leading to the development of nutrient efficient rice varieties.

The image

The image depicts the simulated root structure of rice at 30 days after germination in a virtual soil column. The colour of the root segment represents different root classes. The roots are mirrored back in the column to depict a field-like root distribution. This model was developed and simulated using the open sourced functional-structural plant modelling platform, OpenSim-Root. The simulated root structure was visualised using a freeware application named ParaView.

© Springer Nature Switzerland AG 2020
F. Matthäus et al. (eds.), *The Art of Theoretical Biology*, https://doi.org/10.1007/978-3-030-33471-0_11

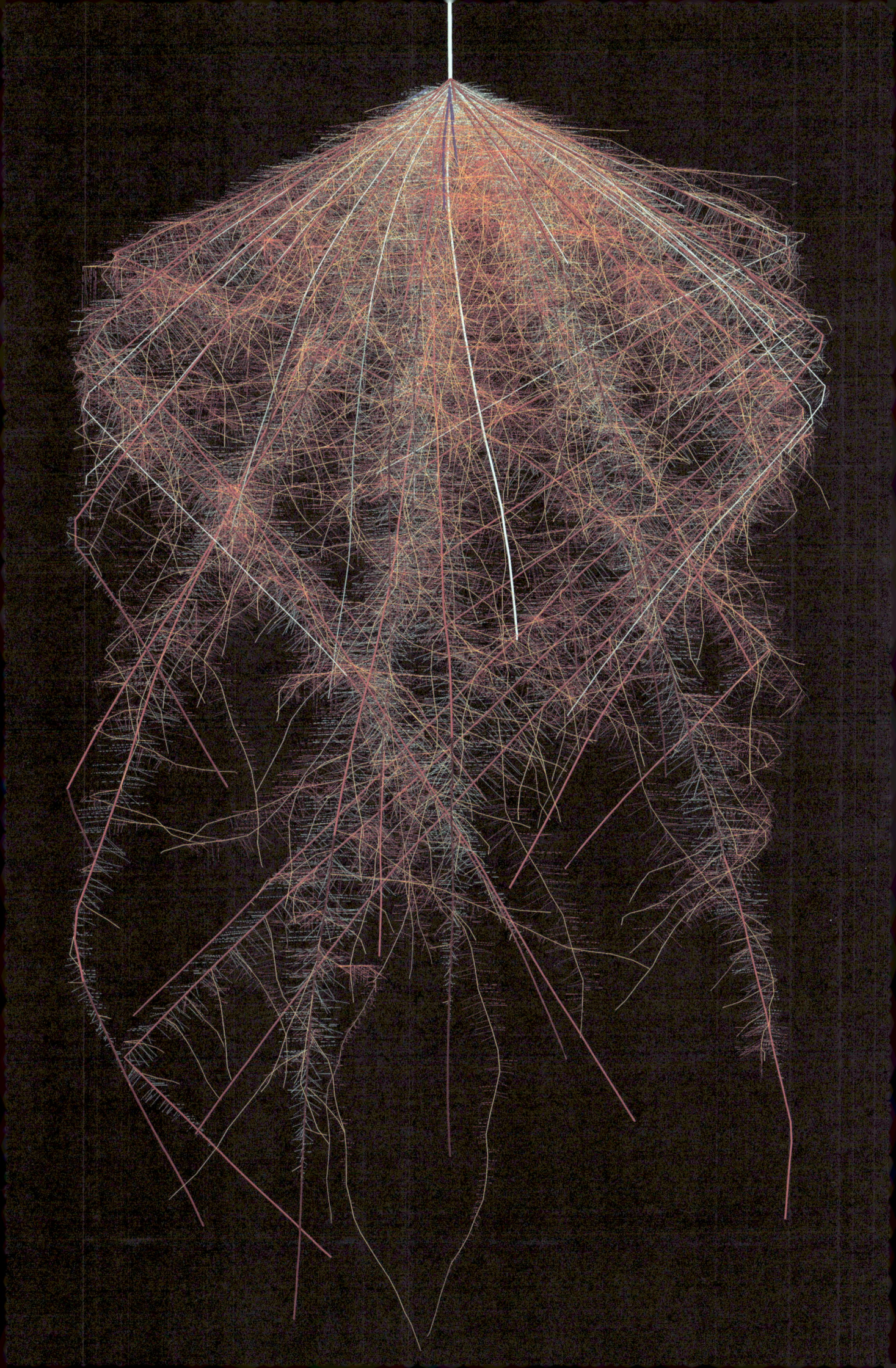

The Beauty of a Beast

By Aenne Oelker, Thomas Horger & Christina Kuttler

The research story

The bacterium Staphylococcus aureus is an important human pathogen, which plays a crucial role in a large number of hospital-acquired infections. We study pattern formation in colonies of this bacterium under laboratory conditions. The patterns formed by colonies of Staphylococcus aureus show specific characteristics for different bacterial strains and mutants. Our aim is to use this knowledge to determine the bacterial mutant types in a cost- and time-efficient way. To this end, we have developed a system of five coupled reaction-diffusion equations describing the densities of replicative and non-replicative bacteria as well as the concentrations of nutrients, quorum sensing signaling molecules and biofilm in the colony. In this context, quorum sensing represents the ability of the replicative bacteria to communicate with each other in order to coordinate their behavior to achieve the best outcome for the survival of the entire population. Furthermore, the biofilm represents the environment in which the colony lives and grows.

The image

To generate the displayed images, time-adaptive numerical finite element simulations of the partial differential equation system have been performed in Matlab. The images display the densities of replicative and non-replicative bacteria as well as the concentration of biofilm, where the color scheme has been adjusted to reflect the single components.

References

[1] García-Betancur JC, Goñi-Moreno A, Horger T, Schott M, Sharan M, Eikmeier J, Wohlmuth B, Zernecke A, Ohlsen K, Kuttler C, Lopez D, Cell differentiation defines acute and chronic infection cell types in Staphylococcus aureus, eLife 6:e28023, 2017.

[2] Horger T, Kuttler C, Wohlmuth B, Zhigun A, Analysis of a bacterial model with nutrient-dependent degenerate diffusion, Mathematical Methods in the Applied Sciences 38(17):3851–3865, 2014.

© Springer Nature Switzerland AG 2020
F. Matthäus et al. (eds.), *The Art of Theoretical Biology*, https://doi.org/10.1007/978-3-030-33471-0_12

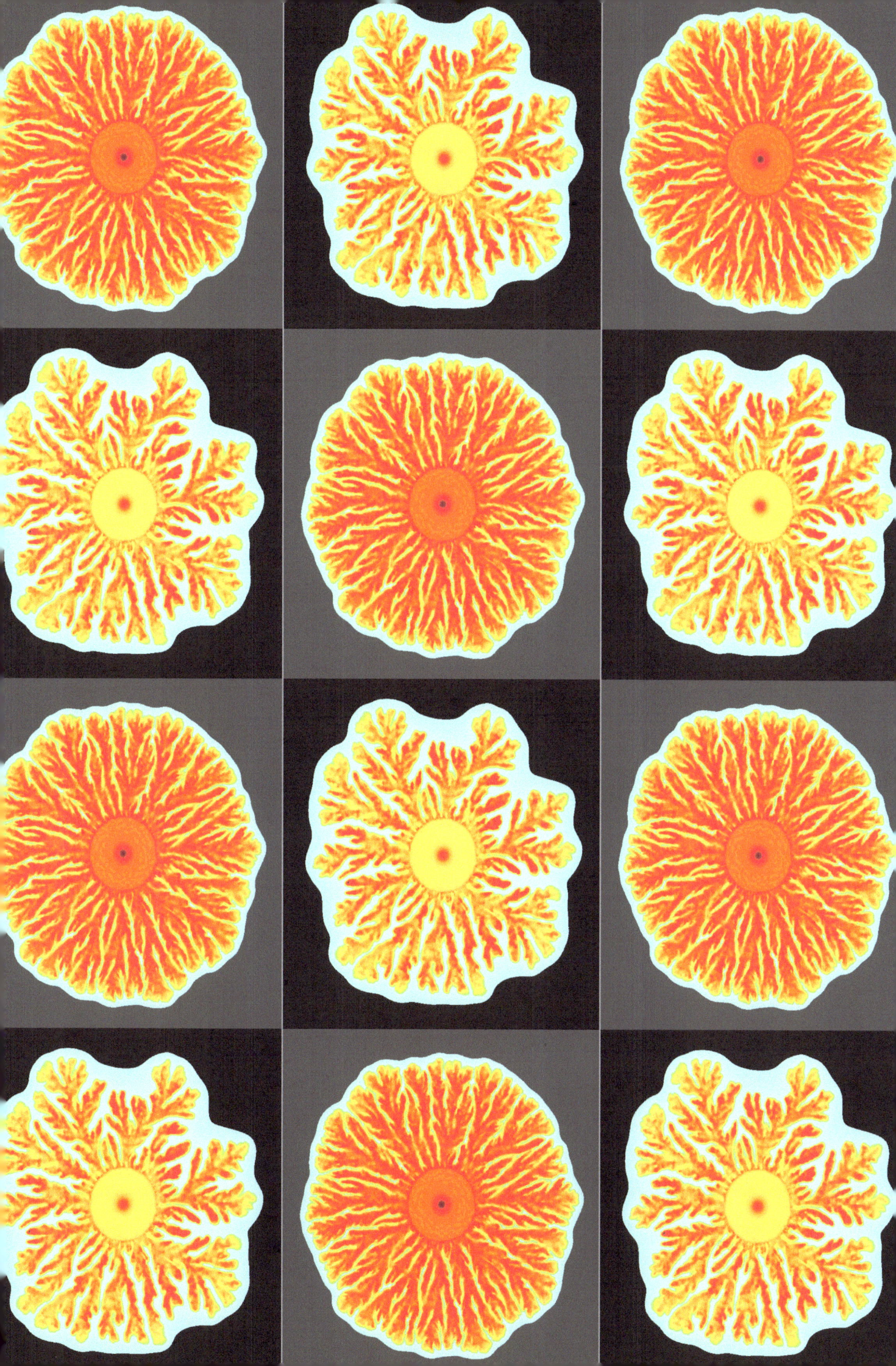

The Ghost

By Franziska Matthäus, Damian Stichel & Kai Breuhahn

The research story

Lung cancer is one of the most serious types of cancer, with high incidence rate and high mortality. The reason for the severity of the disease is the so-called early spread – at the time the disease is diagnosed 40% of patients have already metastases. One research topic is therefore the ability of the cells to detach from the original tumour and invade other tissues. To this end, cell lines – suspensions of cultured immortalized cells – are used for experimental studies. Cell lines originating from lung cancers can exhibit common characteristics of epithelial tissue, which is built from cells that are always in close contact with their direct cell neighbours.

The image

Starting point for this image were experimental data in the form of time-lapse images of collectively migrating cells on a surface. In order to quantify the motion characteristics of the cells and the interaction between the cell neighbors, a correlation analysis, called particle image velocimetry, was carried out. The method yields the average direction and speed of the cells for any given area and every time point. For certain experimental conditions we observed that the cells in the center of the observation area formed a large vortex. We chose one of these experiments and computed streamlines based on the local velocities. The image depicts these streamlines, color-coding the local speeds.

References

[1] Stichel D, Middleton AM, Müller BF, Depner S, Klingmüller U, Breuhahn K, Matthäus F, An individual-based model for the collective cancer cell migration explains speed dynamics and phenotype variability in response to growth factors, NPJ Systems Biology and Applications 3, Art. no. 5, 2017.

[2] Müller B, Bovet M, Yin Y, Stichel D, Malz M, Middleton AM, Ehemann V, Schmitt J, Muley T, Meister M, Herpel E, Singer S, Warth A, Schirmacher P, Drasdo D, Matthäus F, Breuhahn K, Concomitant expression of far upstream element (FUSE) binding protein (FBP) interacting repressor (FIR) and its splice variants induce migration and invasion of non-small cell lung cancer (NSCLC) cells, J. Pathol. 237:390–401, 2015.

F. Matthäus et al. (eds.), *The Art of Theoretical Biology*, https://doi.org/10.1007/978-3-030-33471-0_13

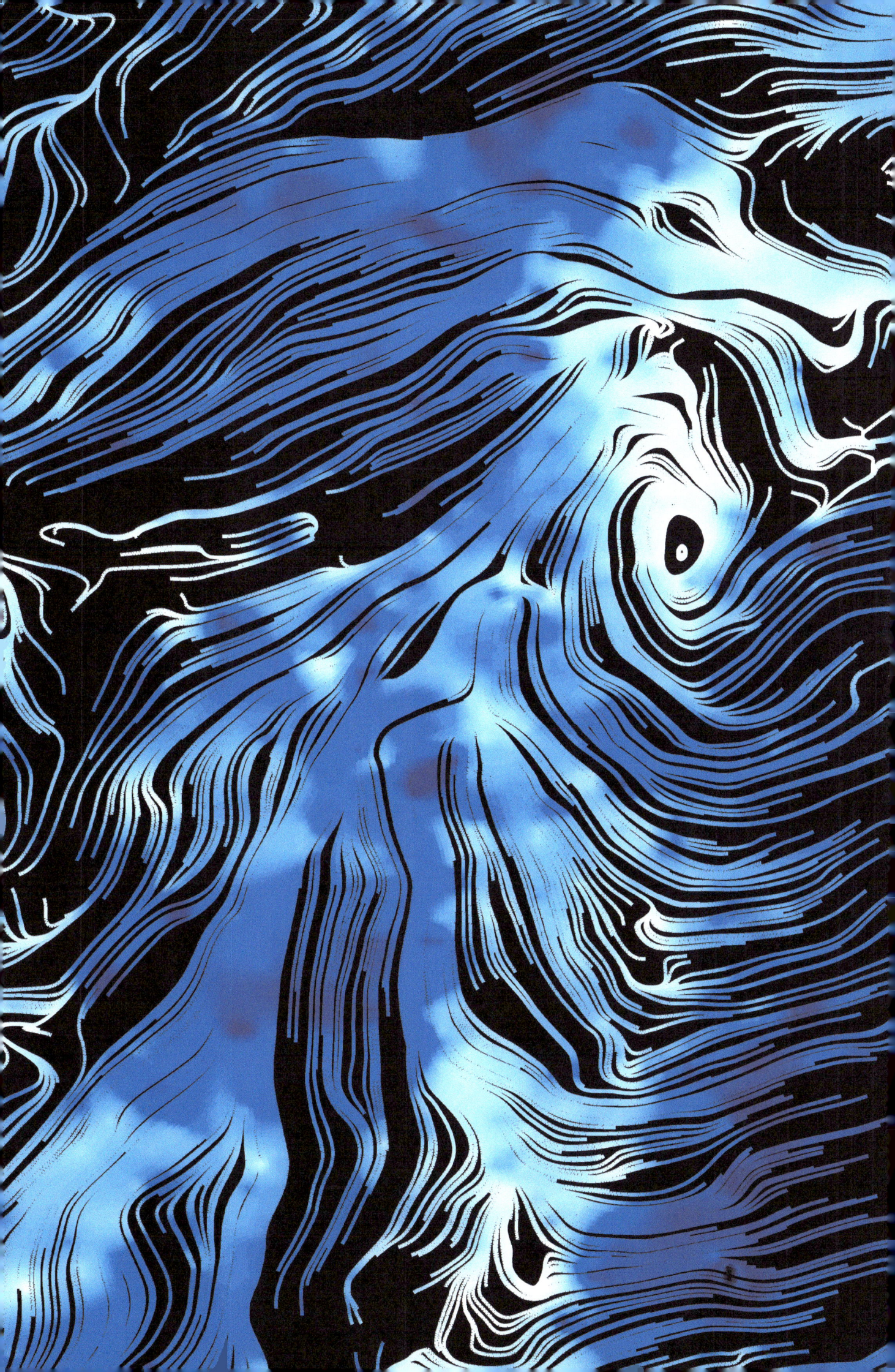

Lymph Node Landscapes

By Hendrik Schäfer, Martin-Leo Hansmann & Ina Koch

The research story

In routine pathology, immunohistological images are a standard tool for diagnostics. Tissue sections are stained highlighting specific parts or cells for visual inspection via light microscopy. Modern whole slide scanners can digitize such tissue samples in high resolution and allow the application of computer-aided methods. We examined classical Hodgkin lymphoma tissue sections. Hodgkin lymphoma is a malignancy of the lymph system and has some unusual characteristics. In contrast to other, solid tumors, only one percent of the tumor tissue is made of malignant cells, surrounded by a heterogeneous environment of immune cells. The malignant cells are called Hodgkin and Reed-Sternberg (HRS) cells. The immuno staining binds to CD30, a protein expressed by HRS cells. Thus, malignant cells appear red in the histological image. A second stain, hematoxylin, binds to negatively charged molecules and color cell nuclei in blue.

The image

We applied an automated imaging pipeline to recognize and chart all HRS cells of the whole slide images [1]. A unit disk graph models the cell distribution in the image. Nodes represent the malignant cells, arcs define the local neighborhood of an HRS cell dependent on the spatial distances. The colored bubbles mark communities computed by clique percolation. Graph theoretically, they are groups of nodes that are highly connected. The image was generated by merging the original histological image with two overlays generated by Impro, our in-house image processing software. The automated object detection yields additional objective data about the cell distribution and morphology and may support the diagnosis of pathologists in future.

Reference

[1]   Schäfer H, Schäfer T, Ackermann J, Dichter N, Döring C, Hartmann S, Hansmann M-L, Koch I, CD30 cell graphs of Hodgkin lymphoma are not scale-free – an image analysis approach. Bioinformatics 32(1):122–129, 2016.

© Springer Nature Switzerland AG 2020
F. Matthäus et al. (eds.), *The Art of Theoretical Biology*, https://doi.org/10.1007/978-3-030-33471-0_14

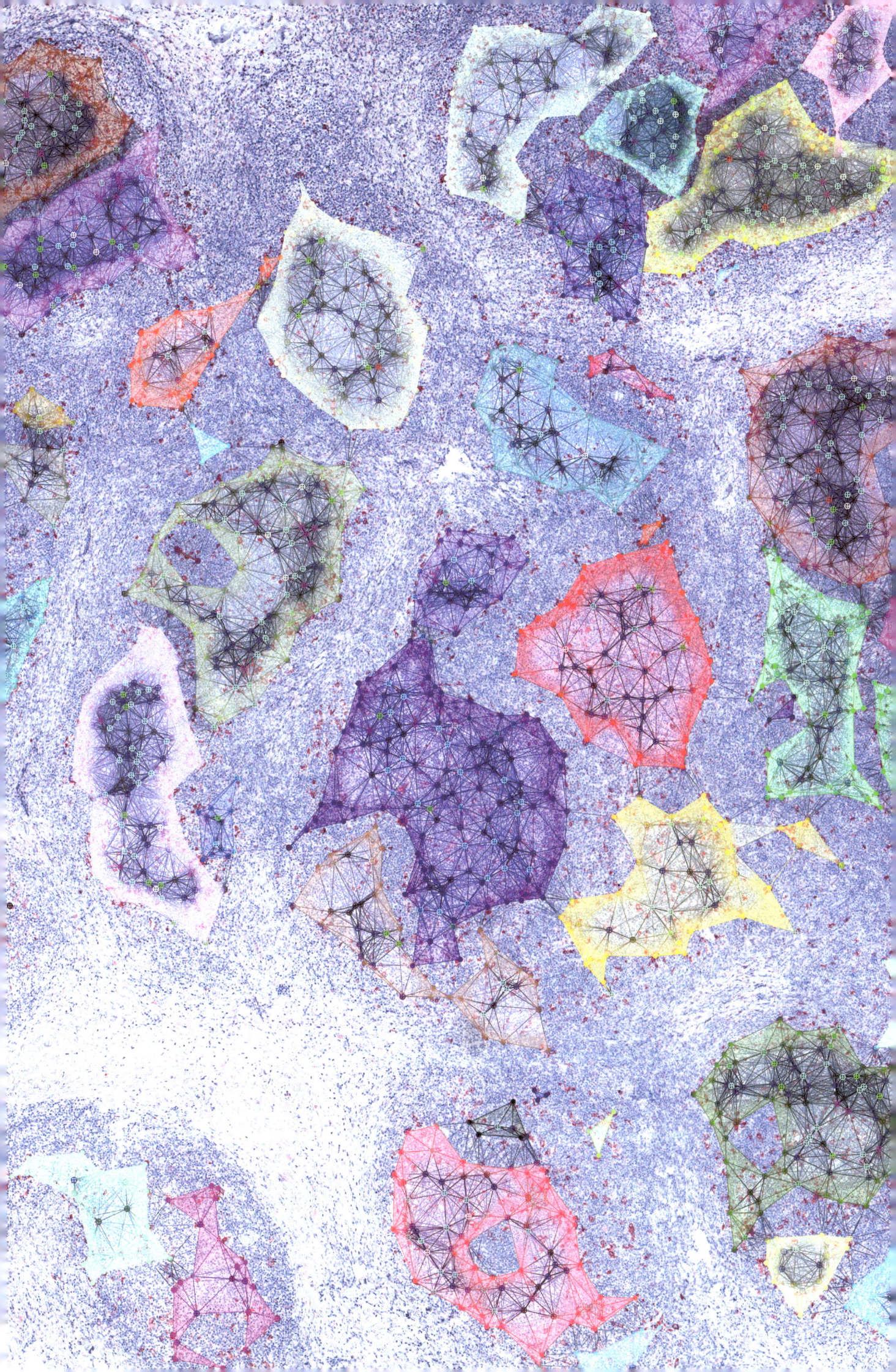

Breezing Drops

By Moritz Mercker & Anna Marciniak-Czochra

The research story

Biological membranes define a mechanical boundary of cells and of substructures inside cells. They provide environments specialized for certain chemical or mechanical processes, and are involved in various healthy and pathogenic cellular activities. A fascinating feature of biological membranes is their ambivalent physical behaviour. On the one hand, with respect to deformations, membranes behave like an elastic body. Frequent deformations are e.g. budded structures, where small vesicles emerge from constrictions of the main membrane. On the other hand, molecules such as lipids can move freely within the presumably deformed membrane surface. From this perspective, membranes behave like a two-dimensional fluid. Complex membrane behaviour and pattern formation arise from the interplay between membrane molecules locally deforming the membrane, and membrane deformations influencing the lateral distribution of membrane molecules in turn. Even simple interplays of this kind can lead to complex chemical and mechanical patterns, so that we need to use mathematical models and simulations to study such processes.

The image

The image shows an assembly of membrane patterns resulting from the interplay of membrane molecules. In the foreground one can see budded structures, which are driven just by the saddle-shape of a particle called ESCRT being released into a lipid membrane [1]. The ESCRT machinery plays an important role in various cellular processes, such as virus infections. In the background, there are three membrane vesicles (closed membrane spheres) where the interplay between different molecule shapes and lipid membranes is investigated. It appears that various chemical or mechanical molecule properties can lead to membrane budding [2].

References

[1] Mercker M, Marciniak-Czochra A, Bud neck scaffolding as a possible driving force during ESCRT-induced membrane budding, Biophysical Journal 108: 833-843, 2015.

[2] Mercker M, Marciniak-Czochra A, Richter T, Hartmann D, Modeling and computing of deformation dynamics of inhomogeneous biological surfaces, SIAM J. Appl. Math. 73(5): 1768–1792, 2013.

© Springer Nature Switzerland AG 2020
F. Matthäus et al. (eds.), *The Art of Theoretical Biology*, https://doi.org/10.1007/978-3-030-33471-0_15

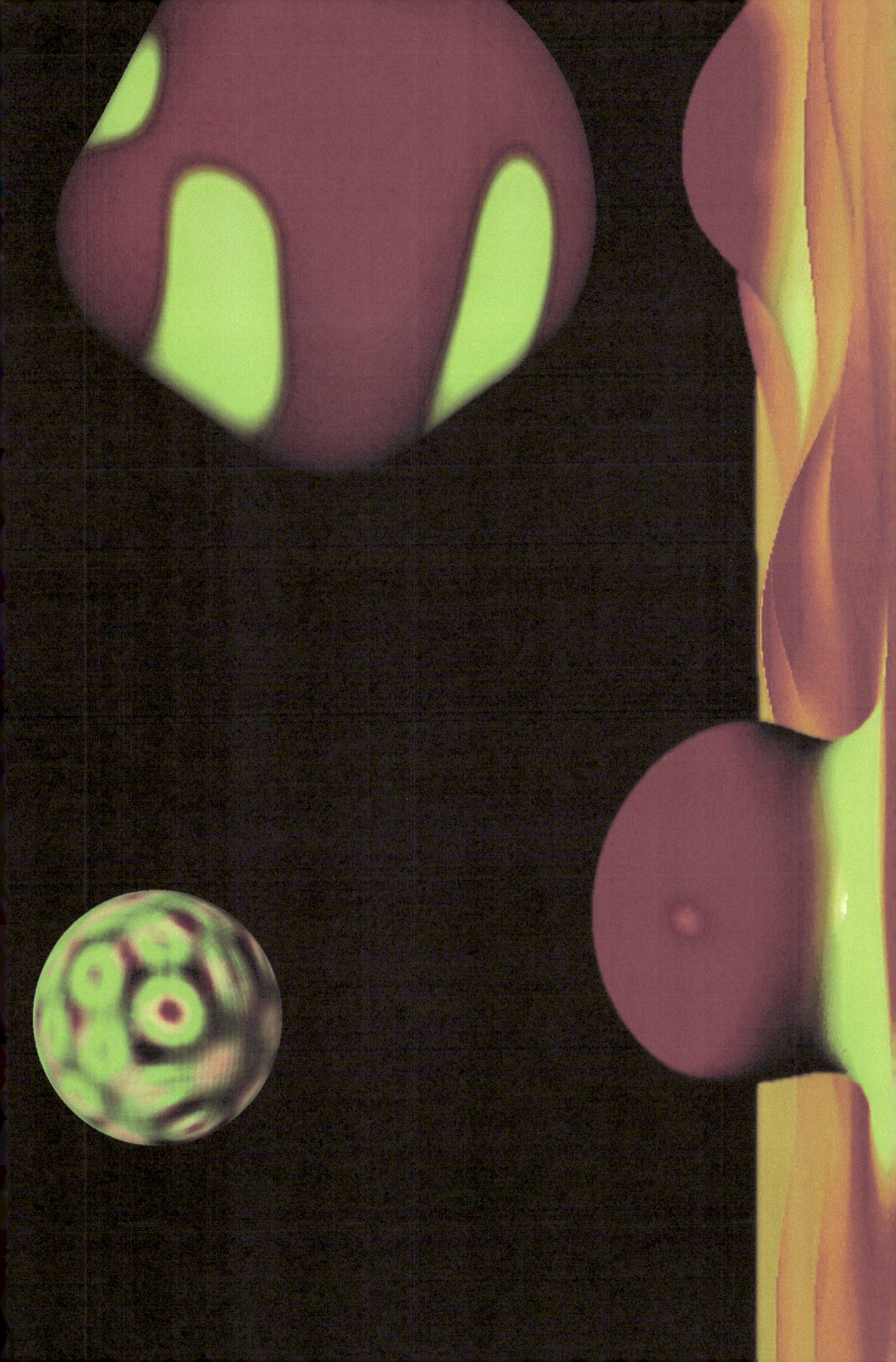

Labyrinths: Exotic Patterns of Cortical Activity

By Aytül Gökçe, Daniele Avitabile & Stephen Coombes

The research story

The cortex of the brain is the seat of human intelligence. It is a thin folded structure and the grey matter on its outside is made up of neuronal cell bodies. Modern neuroimaging methodologies, such as electro- and magneto-encephalography, show us that exotic patterns of spatio-temporal neural activity can form on this excitable tissue in response to sensory input. The spontaneous propagation of waves has also been linked to hallucinations and epileptic seizures. Patterns of activity on the cortical surface can be described using neural field models [1]. These mathematical models of the cortex, that incorporate the brains wiring diagram, are ideally suited to helping us understand the switch between healthy and disturbed dynamic brain states.

The image

The image shows a labyrinthine structure governed by a neural field model, in a patch of flattened cortex with long range axonal connections mediating synaptic interactions. The pattern is seeded from an initial central spot of elevated activity, and over time fingers of activity spread throughout the system. The wake of this spreading activity leaves behind a wonderfully elaborate structure. Here, orange and blue regions represent the excited (high neural activity) and quiescent (low neural activity) states, respectively. Similar patterns have previously been reported and discussed in [2].

References

[1]   Coombes S, beim Graben P, Potthast R, Wright J (Eds.), Neural Fields: Theory and Applications, Springer, 2014.

[2]   Gökçe A, Avitabile D, Coombes S, The dynamics of neural fields on bounded domains: an interface approach for dirichlet boundary conditions, The Journal of Mathematical Neuroscience, 7(1):12, 2017.

© Springer Nature Switzerland AG 2020
F. Matthäus et al. (eds.), *The Art of Theoretical Biology*, https://doi.org/10.1007/978-3-030-33471-0_16

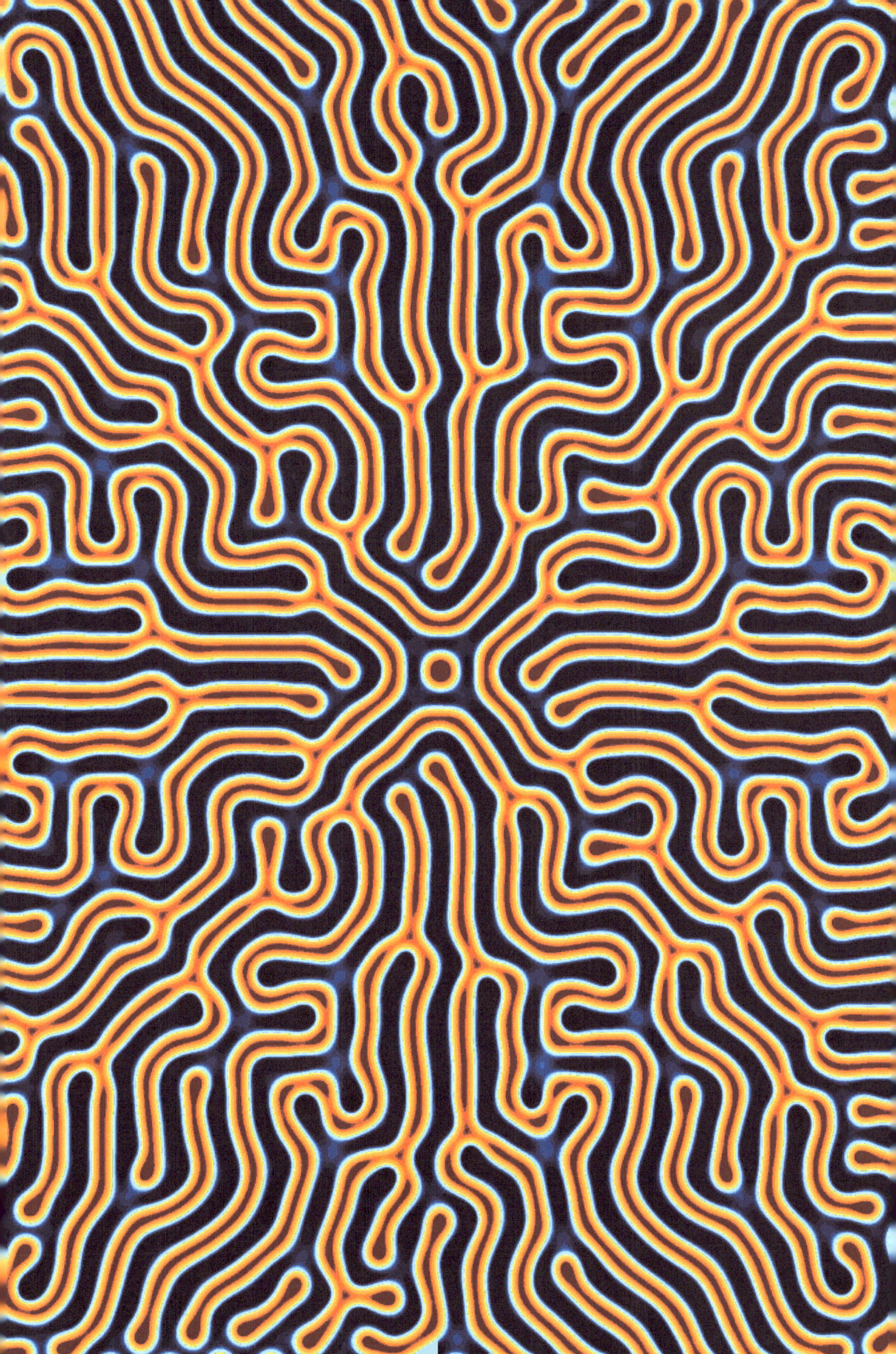

Mammalian Lipidomic Network

By Ferran Casbas & Charlie Hodgman

The research story

Total lipid profiling is being used to find biomarkers for a growing number of disease and mammalian physiological states. However, a more pertinent question is which enzymes have changes in their activity to cause the spectrum of observed lipid changes, because this reveals aspects of the underlying mechanism and could lead to points for therapeutic intervention. This is not straightforward as enzymes of lipid metabolism usually catalyse irreversible reactions of a very large number of potential substrates. To address this question a python script has been written that generates a tripartite graph of 127 enzymes linked to 13934 reactions linked to 7561 lipids. The resulting network underpins a web tool for discovering potential regulators causing lipidome perturbations [1].

The image

This art work shows the entire graph as laid out using Cytoscape. This has been visually enhanced by using a black background and colour-coding nodes by their lipid class, with enzymes and reactions respectively depicted as purple ellipses and light-bordered triangles. Most metabolic networks appear to be a "hairy ball", but this is not the case here, because aspects of lipid metabolism have clustered into particular zones. Orange nodes are diglycerides which go on to produce triglycerides (red), sphingolipids (green) and phospholipids (pink), which are respectively energy sources, nerve-specific cell membrane components and general molecules of lipid-bilayers The pale green nodes on paths leading from the main graph represent synthesis of signalling molecules, including arachinodic acid (involved in inflammation) and cholesterol (the precursor of steroid hormones).

Reference

[1] Casbas-Pinto F, Ravipati S, Barrett DA, Hodgman TC, A methodology for elucidating regulatory mechanisms leading to changes in lipid profiles, Metabolomics 13: 81, 2017.

© Springer Nature Switzerland AG 2020
F. Matthäus et al. (eds.), *The Art of Theoretical Biology*, https://doi.org/10.1007/978-3-030-33471-0_17

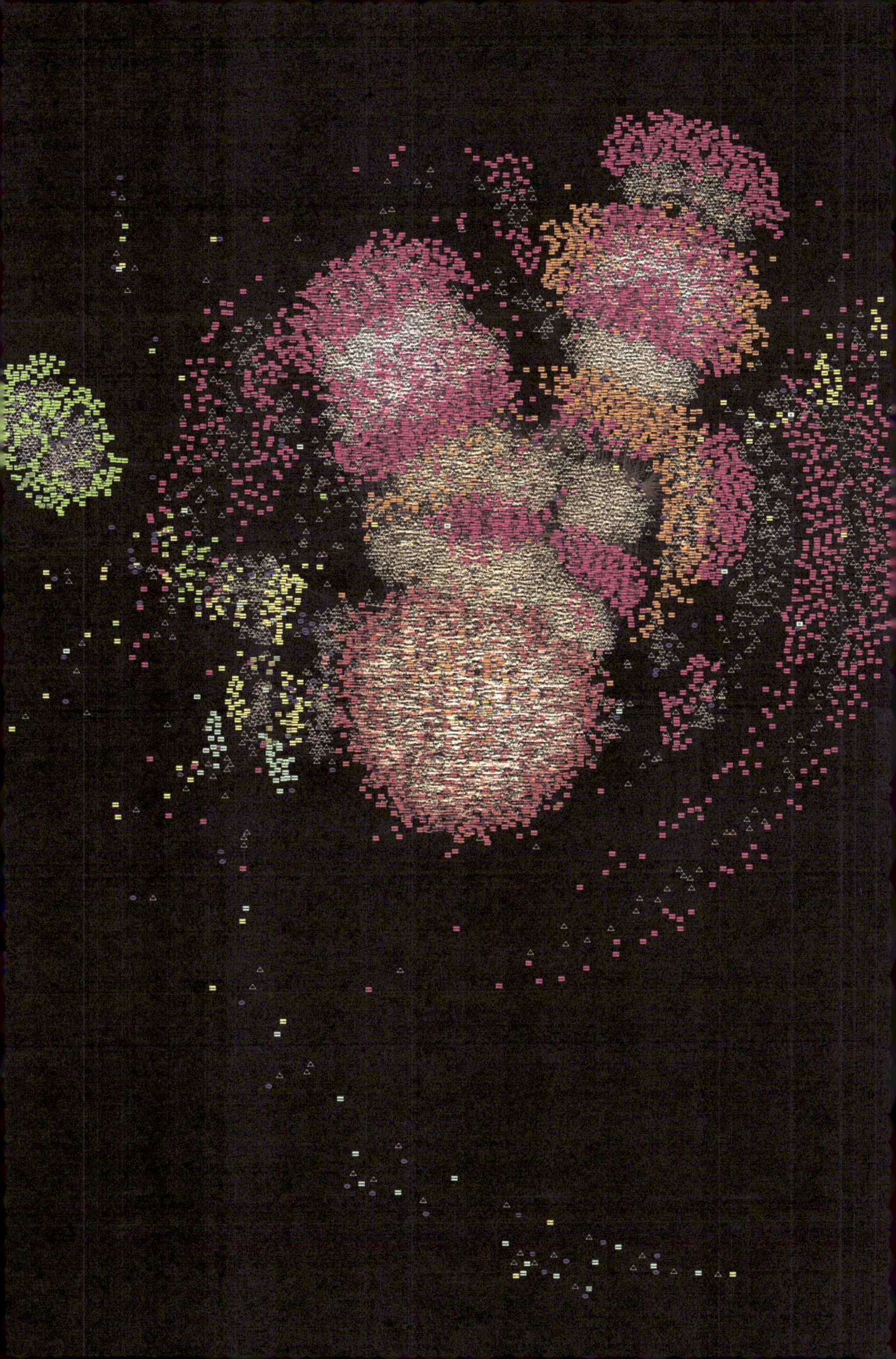

One Step at a Time

By Glenn Webb

The research story

Macrophages are cells that are part of the body's defense in combating infections. One of their important functions is to remove the bacteria we inhale with every breath from the lungs. Macrophages are motile, crawling about on the walls of alveoli (the small air sacs in the lungs) until they locate and eliminate the invader. The macrophage response must be sufficiently rapid and accurate to prevent the proliferation of invading microorganisms. Is random motion sufficient for the macrophages to find their target, or do they adjust their motion by chemotactic sensing? One mathematical model of this process uses random Monte Carlo discrete time steps to model the movements of a macrophage in the alveolus.

The image

In the figure, four Monte Carlo simulations are performed to find the time it takes a macrophage to reach the bacterium (small white circle) at the center of the alveolus (red circle). A single macrophage starts at the right side of the alveolus and moves in uniform length discrete steps (black lines) in a random direction each step. If the macrophage emerges from the circle, it heads radially back toward the center, and this is the way chemotaxis is modeled in the simulations. The number of steps required to reach the target varies in the four simulations. It can be shown that the average number of steps required to reach the target over a large number of simulations is approximately 717.

Reference

[1]   Edelstein-Keshet L, Mathematical Models in Biology, Random House, New York, 1988.

© Springer Nature Switzerland AG 2020
F. Matthäus et al. (eds.), *The Art of Theoretical Biology*, https://doi.org/10.1007/978-3-030-33471-0_18

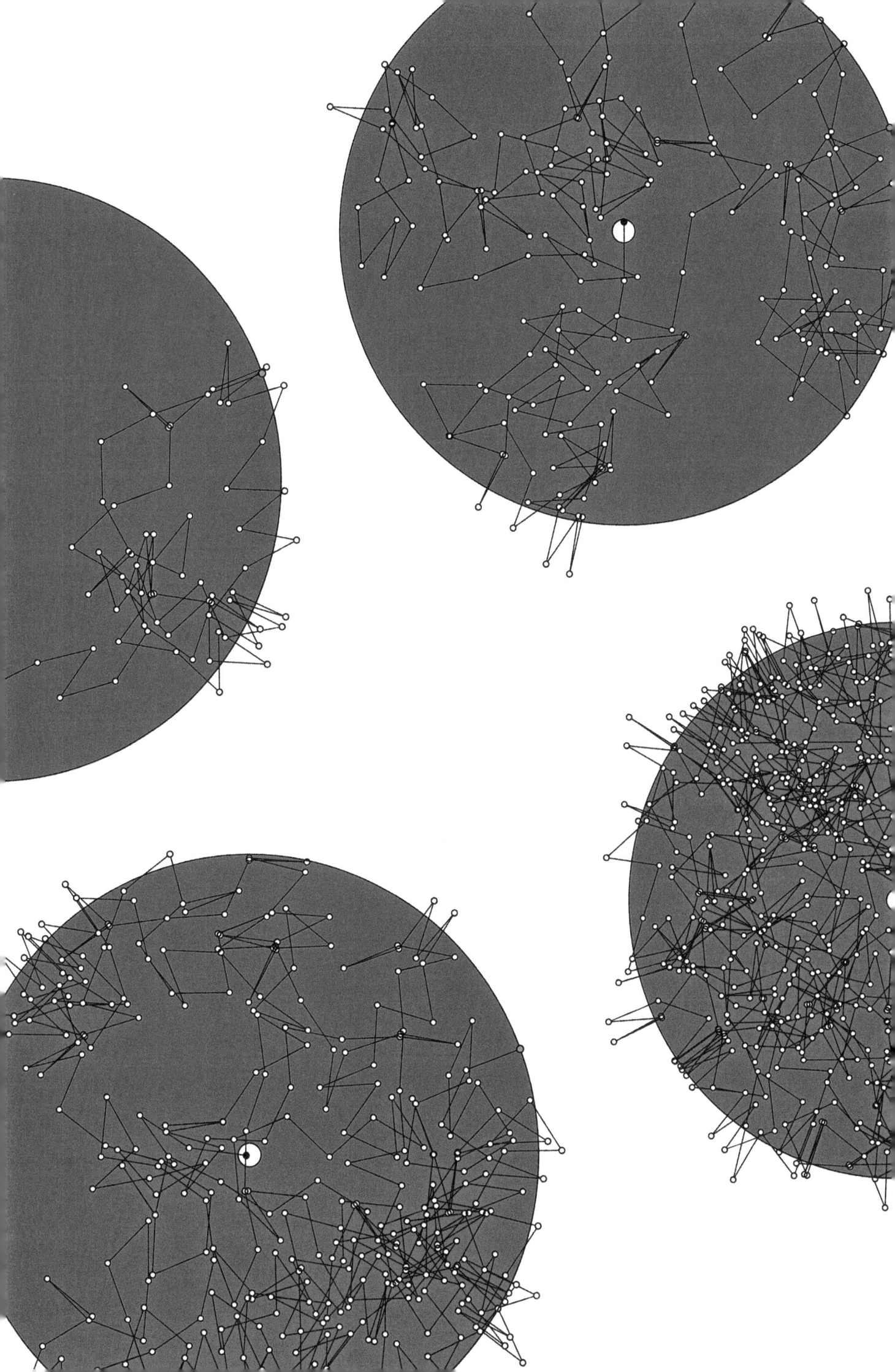

How a Tumor Gets its Spots

By George T Chen, Huaming Yan, Kehui Wang, Robert A Edwards,
Marian L Waterman, Mary Lee, Eric Puttock & John S Lowengrub

The research story

Among the great challenges facing the treatment of cancer is the diversity of cell types within a tumor. Each of these cell types may have different responses to drug treatment, making it necessary to consider using a combination of drugs that target different cell types within the tumor. We focus on the diversity of cell subtypes in colon cancer, which is among the leading causes of cancer-related deaths in the world. Most colon tumors have mutations that aberrantly turn on a critical cell signaling pathway known as Wnt, leading to deregulated, increased cell proliferation rates, among other cancer characteristics. We study Wnt in human colon cancer cells by injecting them into mice and following the activities of Wnt signaling as the injected cells form a vascularized tumor. Even when the cells used to develop these tumors are genetically identical, we observe that fields of cells self-organize via signals to one another to establish a variety of cell subtypes that differ in how they process nutrients.

The image

We observe a striking pattern of heterogeneity in which groups of cells that utilize the resource-intensive process called glycolysis, are encircled by cells using a more resource-efficient mode of energy production called oxidative phosphorylation. Using animal tumors and mathematical simulations, we find that this pattern of cell clusters or "spots" is in part regulated by Wnt signaling. Model simulations suggest that drugs targeting Wnt signaling and glycolysis could serve to effectively treat colon cancer patients by taking advantage of this relationship between Wnt and metabolism.

Reference

[1]   Lee M, Chen GT, Puttock E, Wang K, Edwards RA, Waterman ML, Lowengrub J, Wang K, Waterman ML, Lowengrub J, Mathematical modeling links Wnt signaling to emergent patterns of metabolism in colon cancer, Mol. Syst. Biol. 13: 912, 2017.

© Springer Nature Switzerland AG 2020
F. Matthäus et al. (eds.), *The Art of Theoretical Biology*, https://doi.org/10.1007/978-3-030-33471-0_19

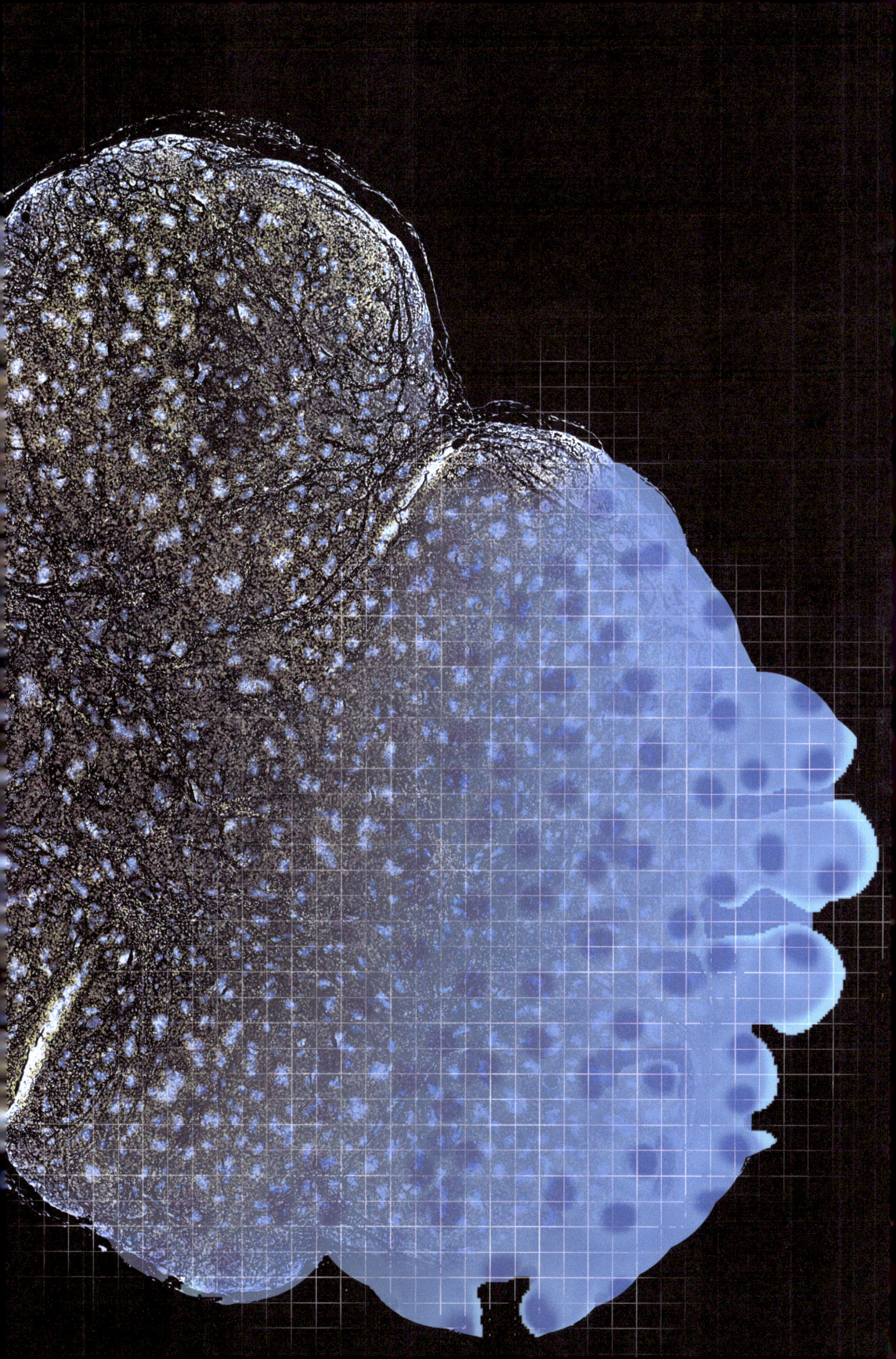

Patchwork Patterns

By Kevin Painter

The research story

In his seminal 1952 paper, "The chemical basis for morphogenesis", British mathematician Alan Turing proposed a novel and elegant theory to explain how the embryo could self-organise and acquire its tissues and organs. In its simplest form it requires just two chemical species, termed morphogens, that react with each other while diffusing through space. Counterintuitively, diffusion acts as the destabiliser, allowing the symmetry of a non-patterned solution to be broken and the chemical concentration to become varied in space. Different chemical levels are proposed to direct cells along separate differentiation pathways, so that heterogeneity can arise from an almost homogeneous starting point. Turing's insights have been used to explain a wide variety of natural patterning phenomena, from how zebras develop their characteristic stripes to the way patchy vegetation can emerge in semi-arid landscapes.

The image

The figure illustrates some of the patterns that can be generated through Turing's mechanism. Computer simulations are used to solve a set of mathematical equations that obey the principals underlying Turing's theory, the Gierer-Meinhardt model. Each frame shows how the chemical concentration varies over a two-dimensional plane under a particular parameter combination, where a randomly-generated colour map is used to indicate different concentration levels. As two of the model parameters are gradually varied, a patchwork of patterns is created, ranging from uniformity to spots to wavy stripes. More than half a century following its original proposal, biologists and mathematicians continue to investigate Turing's theory and are finally beginning to build evidence that it may indeed play a significant role in many processes of embryonic development.

© Springer Nature Switzerland AG 2020
F. Matthäus et al. (eds.), *The Art of Theoretical Biology*, https://doi.org/10.1007/978-3-030-33471-0_20

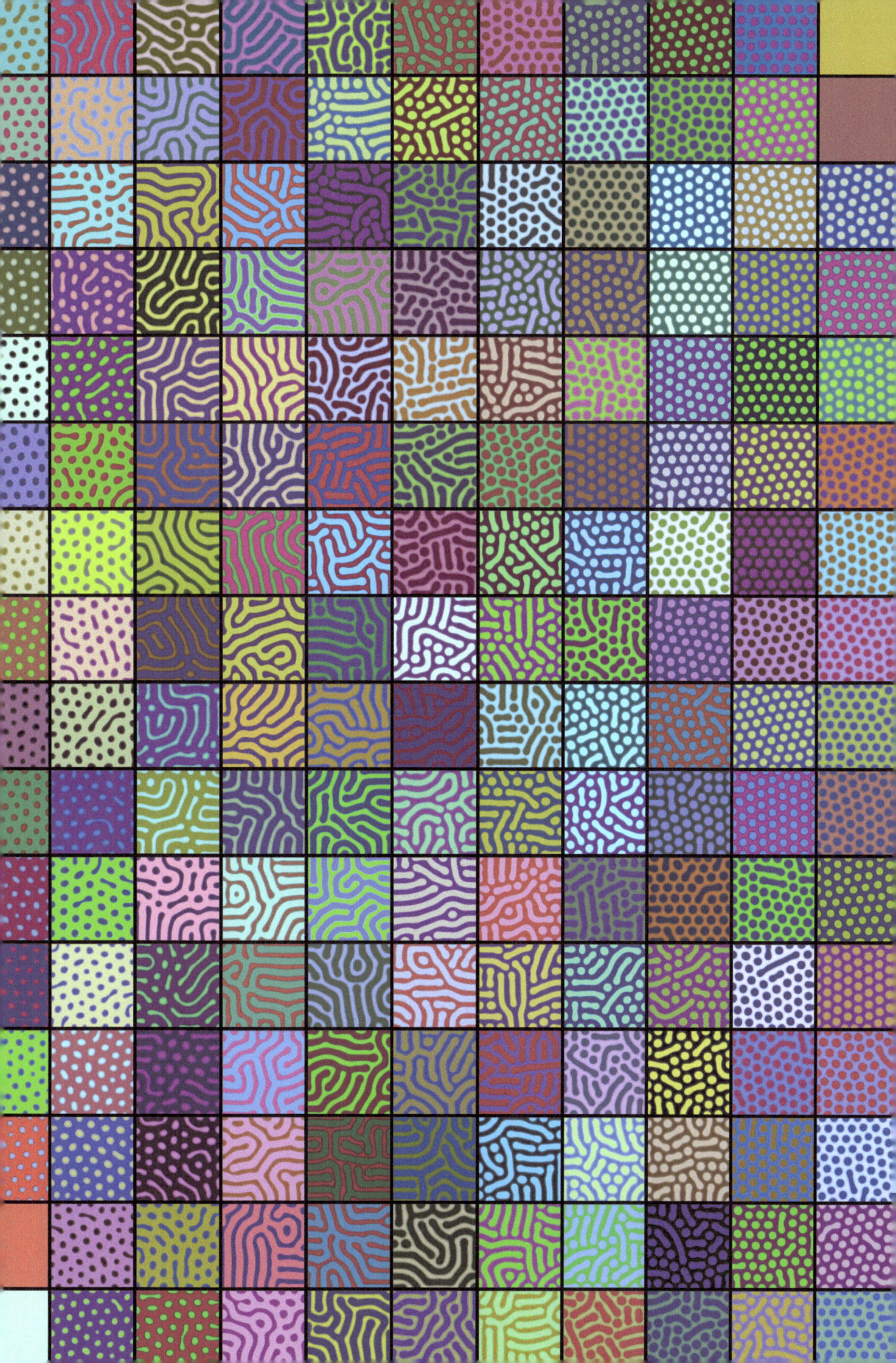

Cancer Warfare

By Randy Heiland, Samuel H. Friedman, Ahmadreza Ghaffarizadeh,
Shannon M. Mumenthaler & Paul Macklin

The research story

Immune cells live in a delicate truce with the rest of the body. When cells behave, the immune system lies dormant. When cells divide and invade spaces where they don't belong, they attract the attention of immune cells to quell the insurrection. Normal cells have "transponders" on their surfaces that help immune cells recognize them as friendly. Cancer cells have abnormal transponders that drive immune cells to attack and clear out the cancer cells. We built a computer model to understand how immune cells track down and destroy cancer cells, and why they often fail to defeat cancer.

The image

In the picture, immune cells (red) follow chemical trails to find cancer cells (yellow and blue). Whenever an immune cell bumps into a cancer cell, it latches on, tests its "transponder" (yellow cells are abnormal), and tries to kill it. It either succeeds, or it stays attached and tries until giving up. At first, this is very successful, and the tumor shrinks. But immune cells can overlook cancer cells while following the chemical trail, trapping themselves in a useless clump. This allows the cancer to regrow, often with mutations to ward off further immune attacks. We used PhysiCell [1] (a free simulation toolkit we created) to build a computer model of each immune and cancer cell. We wrote cell rules that reflect our understanding of how immune cells migrate and attack cancer cells. Supercomputers can run hundreds of simulations to find better immune cell battle tactics [2].

References

[1] Ghaffarizadeh A, Heiland R, Friedman SH, Mumenthaler SM, Macklin P, PhysiCell: an open source physics-based cell simulator for 3-D multicellular systems, PLoS Comput. Biol. 14(2):e1005991, 2018.

[2] Ozik J, Collier N, Wozniak JM, Macal C, Cockrell C, Friedman SH, Ghaffarizadeh A, Heiland R, An G, Macklin P, High-throughput cancer hypothesis testing with an integrated PhysiCell-EMEWS workflow, BMC Bioinformatics 19(Suppl 18): 483, 2018.

© Springer Nature Switzerland AG 2020
F. Matthäus et al. (eds.), *The Art of Theoretical Biology*, https://doi.org/10.1007/978-3-030-33471-0_21

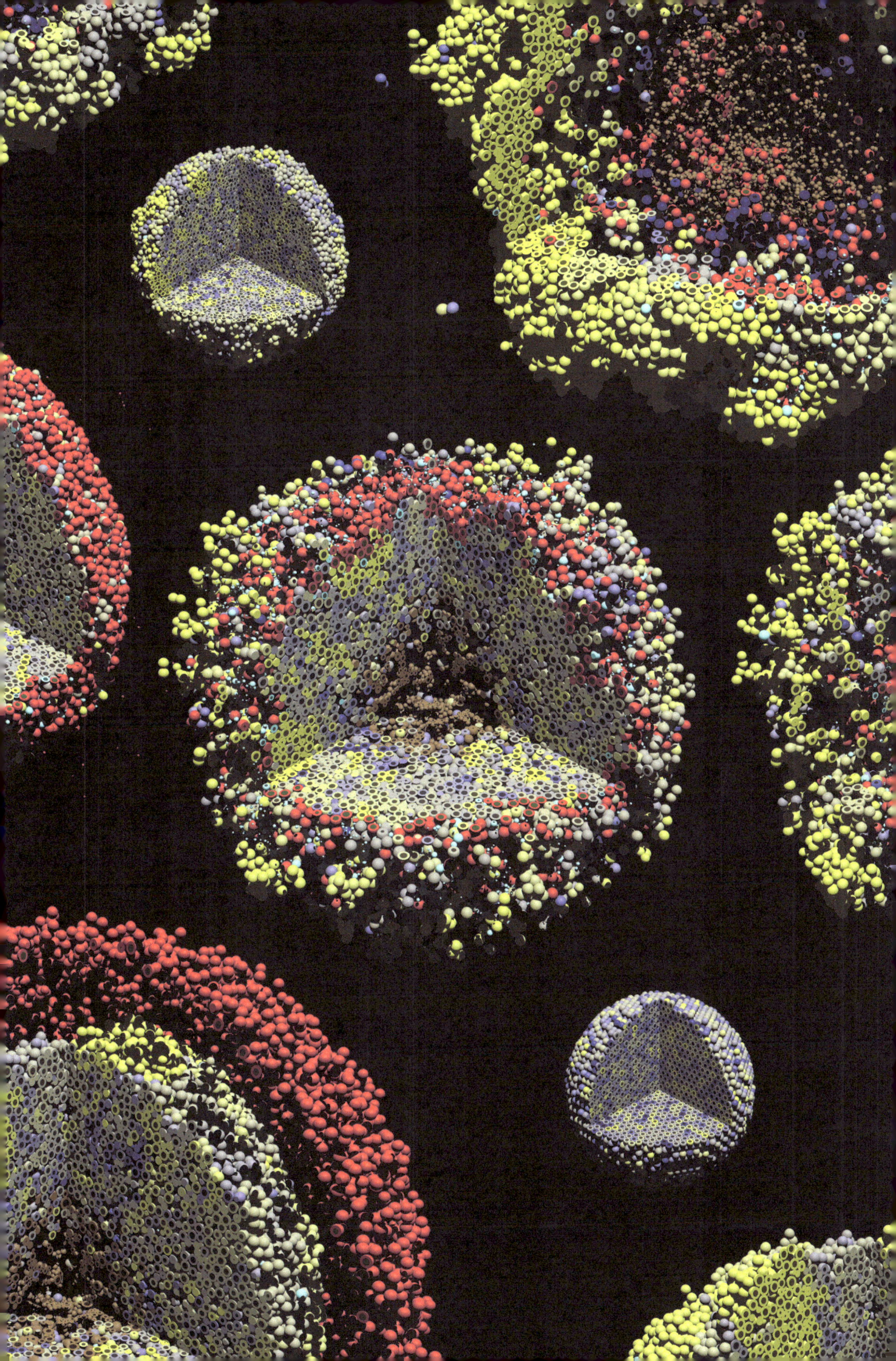

Collective Decision Making

By Wim Hordijk

The research story

A common property of biological systems is the interaction between many individual parts, resulting in their ability to collectively solve problems. For example, billions of neurons in the human brain, all connected together, enable us to find our way home, or recognize a familiar face in a crowd. Similarly, ants in an ant colony interact through chemical signals, and manage to build intricate nests and social structures. However, most of these interactions happen only locally, between nearby neurons or ants. Yet collectively they solve problems, or make decisions, at a system-wide level. We use cellular automata to gain more insight into this process. Cellular automata are simple models of such systems with only local interactions, but they show surprising abilities to self-organize and make collective decisions.

The image

Here we gave a cellular automaton the following task: decide whether there are more white or black cells in the first (left) line of the image. As the system progresses from left to right, line by line, the local interactions between individual cells give rise to competing patterns of white and black, with eventually one of them winning, thus resulting in a collective decision (black in the top image, white in the bottom one).

Reference

[1]   W. Hordijk, The EvCA project: A brief history, Complexity 18(5):15–19, 2013.

© Springer Nature Switzerland AG 2020
F. Matthäus et al. (eds.), *The Art of Theoretical Biology*, https://doi.org/10.1007/978-3-030-33471-0_22

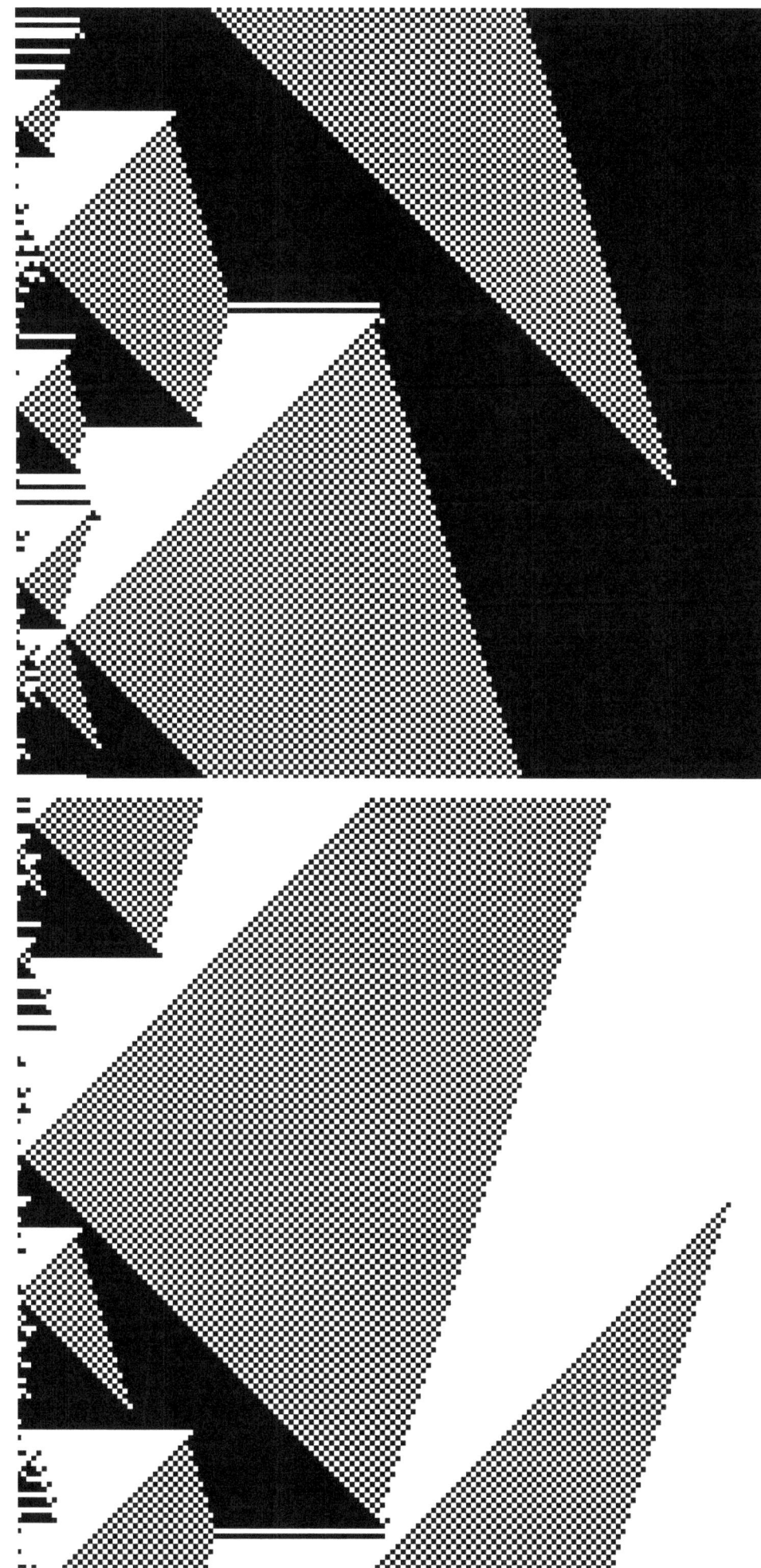

Cell Simulation in Blossom

By Armin Drusko & Franziska Matthäus

The research story

Microscopy has experienced immense advances during the past years. Presently it is possible to observe three-dimensional multicellular systems with high temporal and spatial resolution. 3D life-cell imaging is applied to study the growth and restructuring of artificial tumors, or the development of insect embryos up to the larvae stage. In all of these systems cell migration plays an important role and many studies focus on the question of how cell migration is regulated by chemical and mechanical cues. In order to gain understanding about the cell dynamics in these three-dimensional multicellular systems, we develop and apply image analysis tools for time-lapse microscopy. When the individual cells can be clearly distinguished then the tracking of the paths of single cells is possible. When the cells are dense or the image quality low, a method called particle image velocimetry (PIV) is commonly applied. PIV relies on correlation of image segments and provides a velocity field for every pair of consecutive images.

The image

The image shows the paths of a large number of simulated cells which move collectively in a confined 3D cylindrical area. Movement of the cells underlies mechanical forces like attraction and repulsion, which regulate the cell-to-cell distances. Furthermore, a pre-defined velocity field forces the cells on torus-like paths (down at the outer rim, up in the middle). The colors are coding the direction of motion. The simulation provided artificial data to test an extension of PIV to 3D (plus time) time lapse image analysis.

F. Matthäus et al. (eds.), *The Art of Theoretical Biology*, https://doi.org/10.1007/978-3-030-33471-0_23

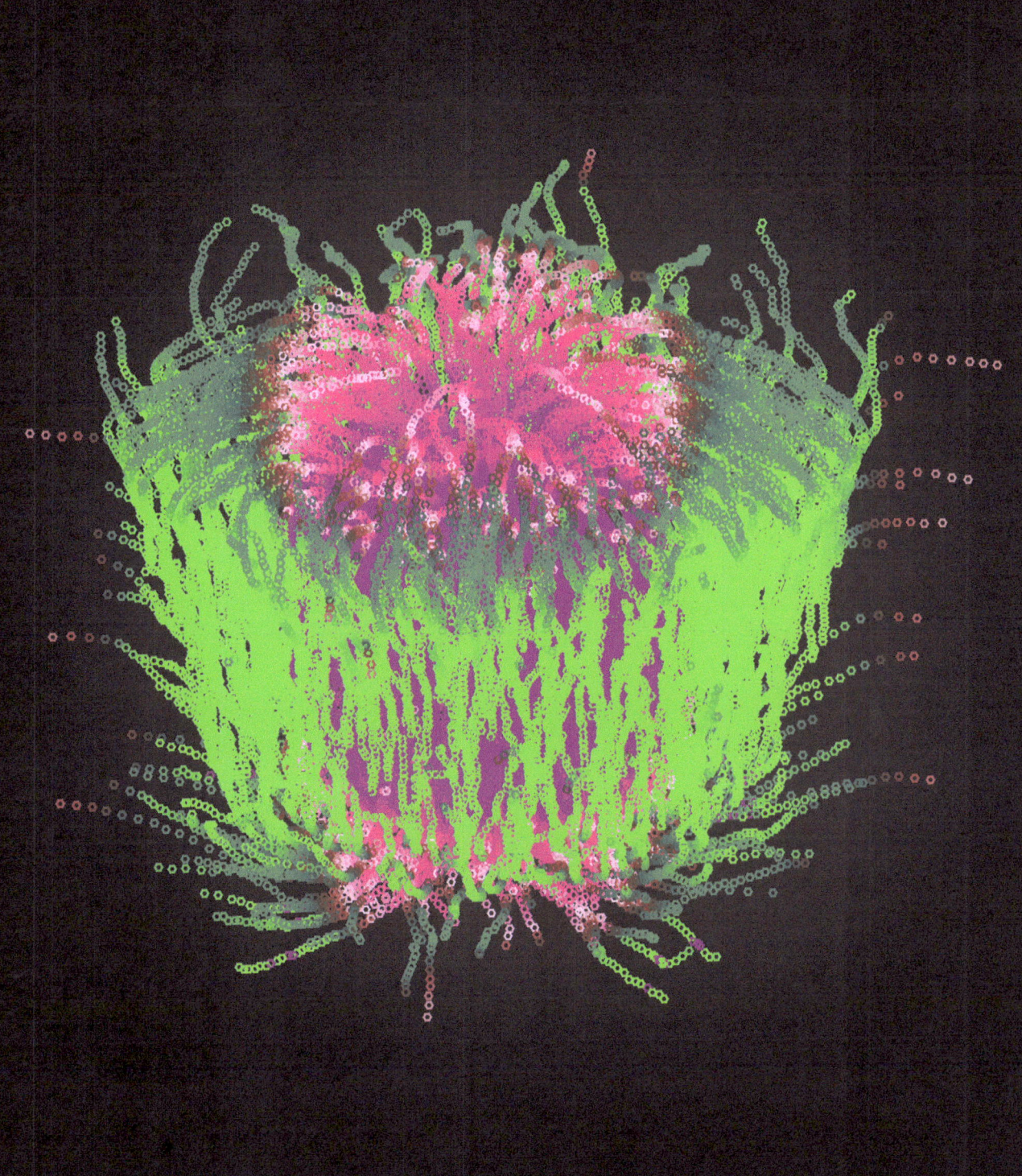

Semblance of Heterogeneity

By Linus Schumacher, Philip Maini & Ruth Baker

The research story

Recently I became interested in heterogeneity in collective cell migration. If some cells in a group move differently, how does that change the overall group motion? For example, some cells can be more invasive in moving through tissues, which can be important in embryo development as well as cancer metastasis. When one measures how different cells in a group react to each other, however, it can be difficult to know how much variability is biologically relevant and has an effect on the collective cell migration, and how much variability is irrelevant noise. This conundrum led me to a mathematical experiment that produced this image.

The image

The image shows simulated tracks of collective cell migration. Each colored line is the trajectory of a different cell. Cells interact by aligning their motion with their immediate neighbours, as well as pulling towards them and pushing away if they get too close – a much simplified representation of cell interactions. In each group in the image, we see different emergent behaviours depending on the strength of alignment relative to pushing and pulling: disordered clumps, ordered structures, and swarm-like motion. In this model, all cells are identical, yet when you statistically analyse their movement, the population can appear heterogeneous [1]. This "Semblance of Heterogeneity" comes about because of the interactions between many cells. With simplified mathematical models like this one we can calculate what we expect to measure in complex biological settings where our intuition fails, and thus guide the interpretation of experiments.

Reference

[1]   Schumacher LJ, Maini PK, Baker RE, Semblance of heterogeneity in collective cell migration, Cell Systems 5(2): 119–127.e1, 2017.

© Springer Nature Switzerland AG 2020
F. Matthäus et al. (eds.), *The Art of Theoretical Biology*, https://doi.org/10.1007/978-3-030-33471-0_24

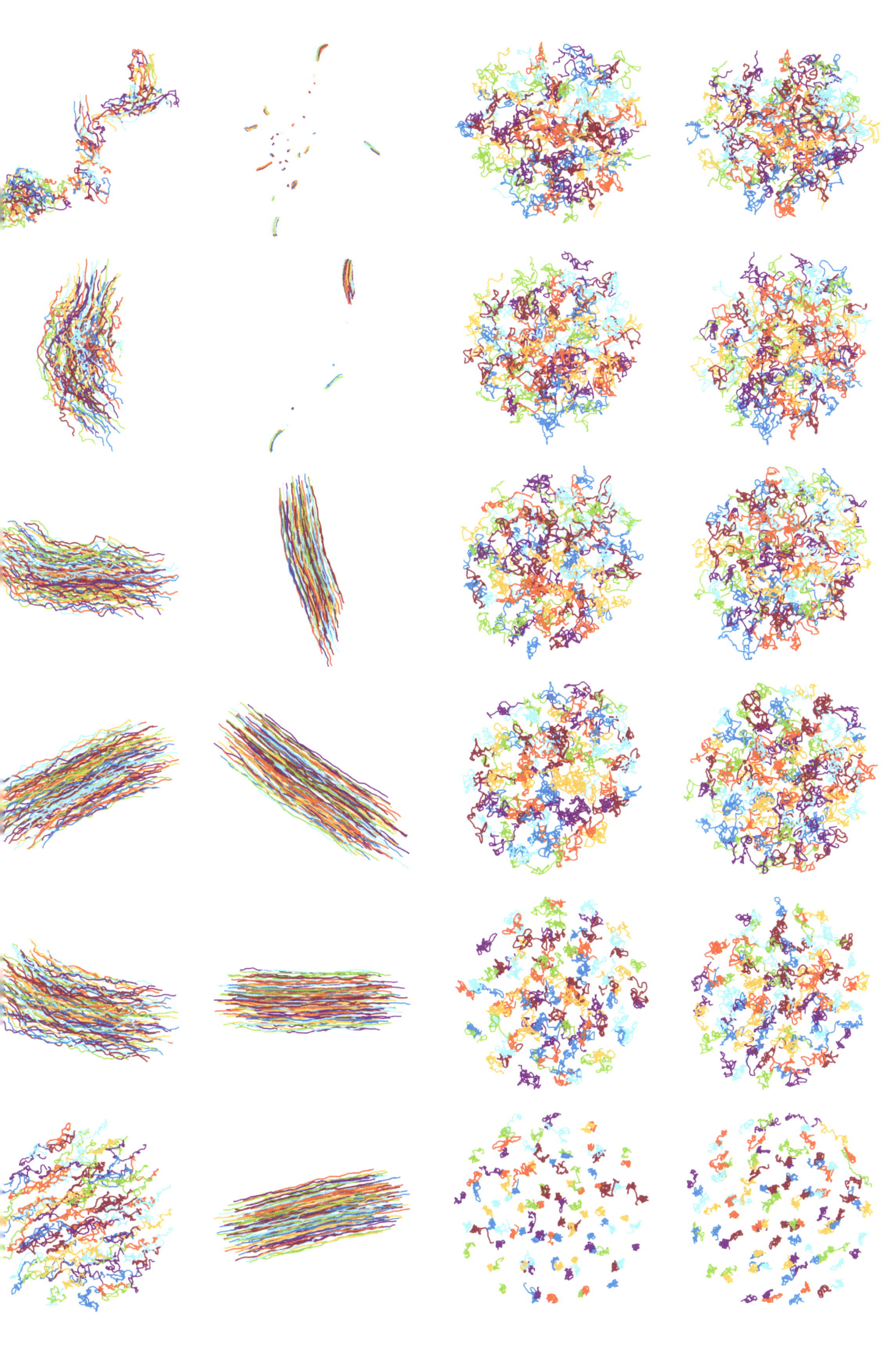

Can we Crack Cancer?

By Sara Hamis & Gibin Powathil

The research story

Contemporary mathematical models of solid tumours constitute an invaluable complement to traditional cancer research [1]. Accordingly, we have developed an in silico framework (a computational framework), capable of simulating solid tumours under multiple conditions, emulating established clinical scenarios. Our framework is based on a mathematical model formulated by biological knowledge, medical experience, experimental observations and mathematical concepts. This tumour model is biologically detailed and mathematically rigorous as we account for the ever-changing tumour on multiple scales. Using our computational framework, multiple aspects of tumour dynamics can be studied, such as tumour growth, drug resistance and intercellular interactions [2]. Importantly, the model can also be used to simulate treatment responses to various combinations of anticancer therapies. These anticancer therapies include traditional treatment strategies such as chemotherapy and radiotherapy as well as novel, pre-clinical therapies currently in development. Simply put, our research premise is that if we can describe cancer using mathematics, then we can predict cancer using mathematics.

The image

The image demonstrates how oxygen levels vary amongst cancer cells in a simulated, cracked open tumour spheroid where the different colours correspond to various oxygen concentrations. Cells with high oxygen concentrations are displayed in warm colours such as yellow and red whilst cells with low oxygen concentrations are displayed in cold colours such as blue. In our simulations, each cell displays individual properties, such as oxygen concentration, that affect how the cell responds to anticancer therapies. The mathematical model is implemented in C++ and visualised using ParaView.

References

[1] Hamis S, Powathil GG, Chaplain MAJ, Blackboard to Bedside: A mathematical modeling bottom-up approach toward personalized cancer treatments, JCO Clin Cancer Inform 3:1–11, 2019.

[2] Hamis S, Nithiarasu P, Powathil GG. What does not kill a tumour may make it stronger: In silico insights into chemotherapeutic drug resistance, J. Theor. Biol. 454:253–267, 2018.

© Springer Nature Switzerland AG 2020
F. Matthäus et al. (eds.), *The Art of Theoretical Biology*, https://doi.org/10.1007/978-3-030-33471-0_25

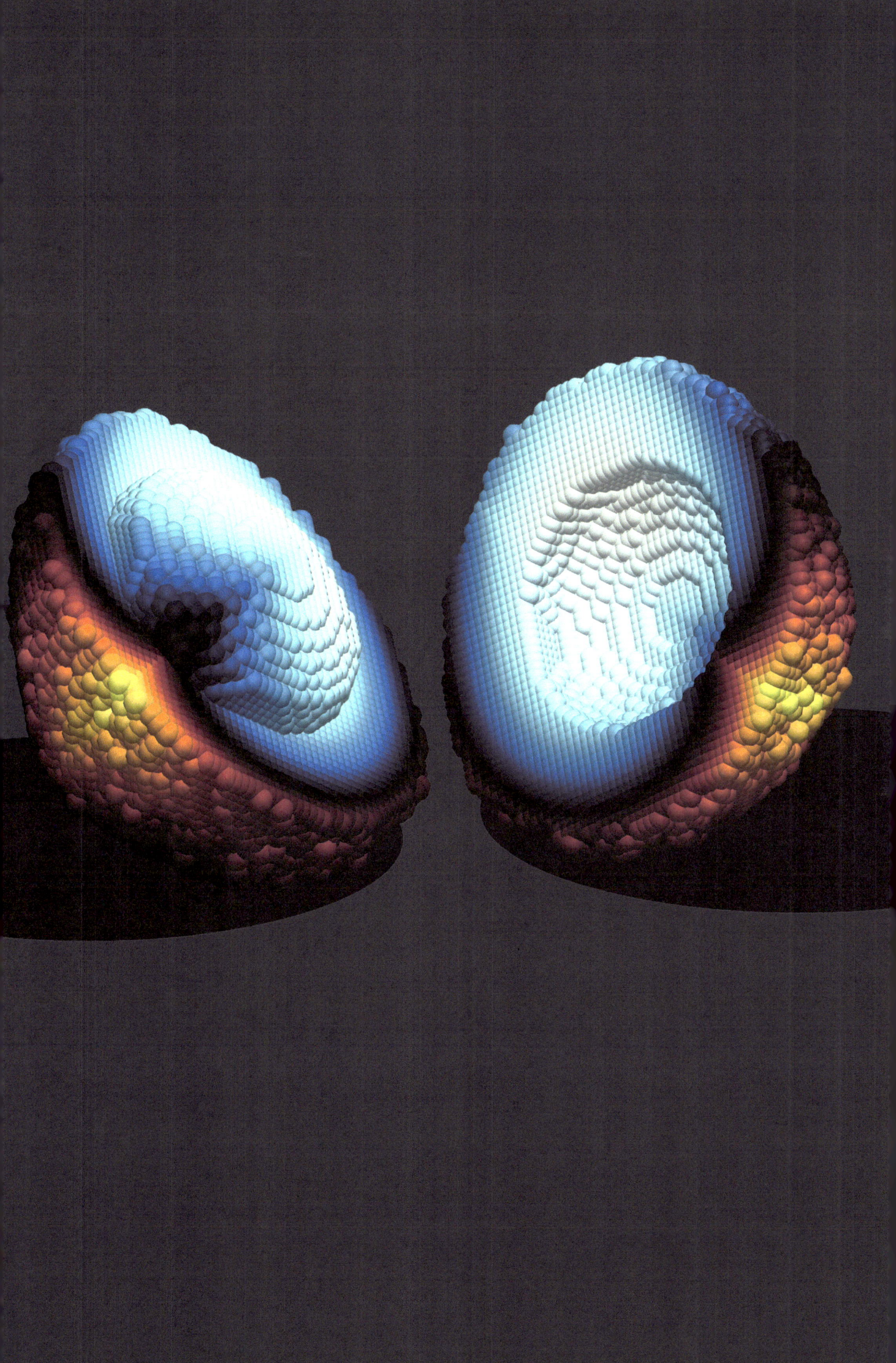

Dance with Predators and Prey

By Yong-Jung Kim

The research story

Sea plankton community produces half of the oxygen and nutrient for animals on earth. Due to such an importance in the ecosystem, the patchiness phenomenon of sea plankton has been one of the key issues in ecology for a long time. The plankton community consists of two groups, phytoplankton and zooplankton, which are often explained as prey and predator species, respectively. Lotka-Volterra equations with diffusion have been a key system to understand the phenomenon. However, the system should be properly modified to produce persistent patterns and various modified equations have been considered. The provided dynamical patterns are obtained after adding an extinction mechanism only. It is not surprising at all that the beautiful patterns of life forms cannot be obtained without the death.

The image

The figure illustrates some of time sequential patterns, which are solutions of Lotka-Volterra equations with diffusion and finite time extinction. If the first stage is finished, time periodic patterns appear repeatedly. These lively patterns look like dancers in a synchronized group dance. An interesting feature is that the background takes a spotlight soon after and a spotlighted dancer becomes a background dancer a moment later. Such a balance makes the life lively. If one insists to stay under a spotlight forever, the dancer will become a cancer and the lively pattern of life will fade away. In the image, there are 16 pairs of prey and predator patterns chosen at the same moment respectively. Can you match all of them? Find more details and examples of patterns in predator-prey equations from http://amath.kaist.ac.kr/predatorprey/.

Reference

[1] Choi J, Kim Y-J, Predator-prey equations with constant harvesting and planting, Journal of Theoretical Biology 458: 47–57, 2018.

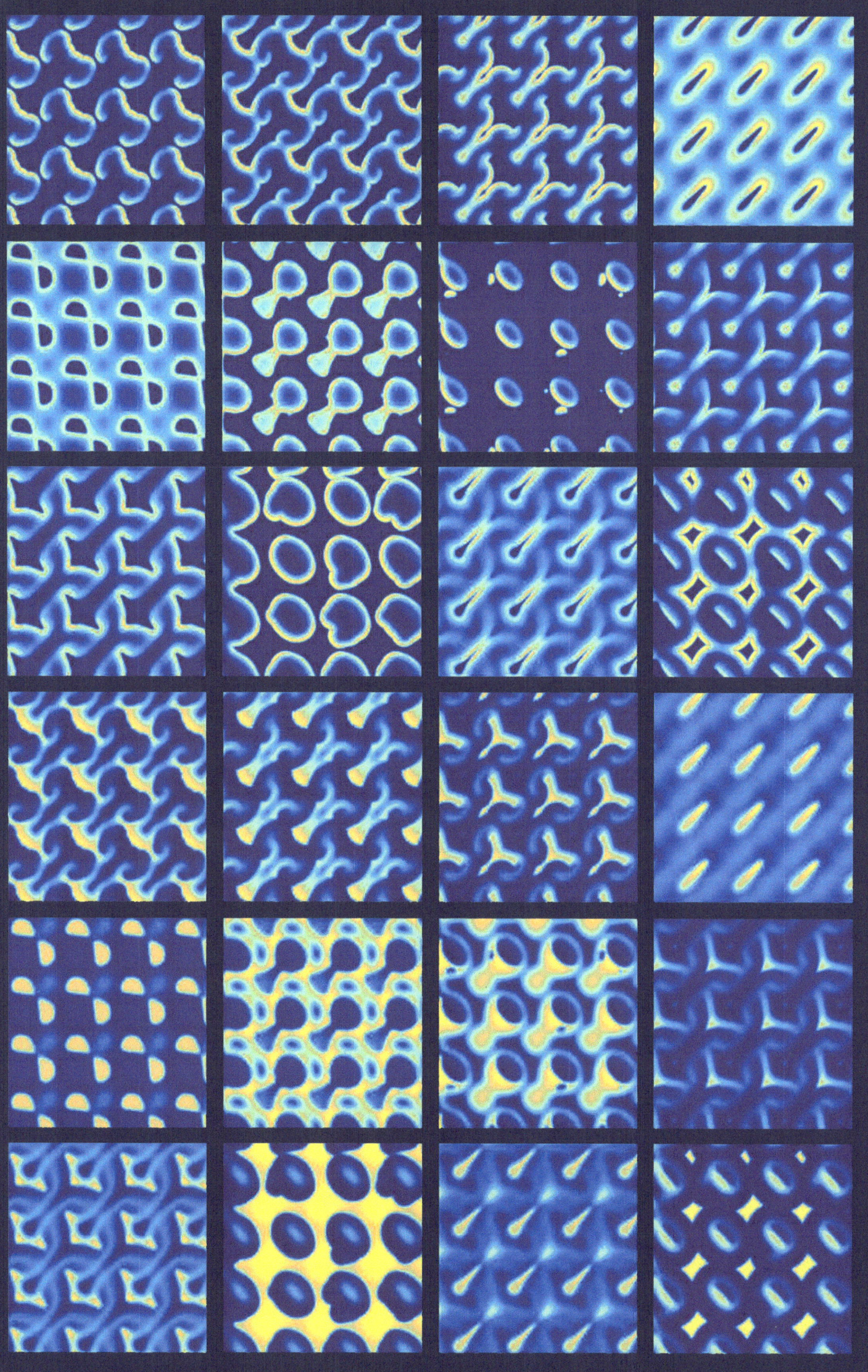

Knitting Proteins

By Santiago Schnell

The research story

Proteins are made of a long sequence of building blocks called amino acids. Proteins fold into a complicated three-dimensional shape using spontaneous origami rules designed by nature. How a protein folds into a specific shape determines how it will act. If protein folding origami fails, proteins remain misfolded and cannot function. Many misfolded proteins can trigger diseases known as protein folding diseases. As a matter of fact, a growing number of aging and degenerative diseases are associated with inappropriate deposition of misfolded proteins into aggregates or fibrils in our cells and tissues. Aggregates or fibrils can assemble through a molecular knitting process, in which misfolded protein molecules bind together in a disorganized structure (aggregates) or regularly organized chain (fibril). The time it takes to complete the molecular knitting process is important to design treatments to inhibit aggregation or fibrillation in protein folding diseases [1].

The image

The image shows a molecular computer model of defective insulin molecules [2] forming several threaded fibrils. These fibrils can form in the skin of patients suffering from Type 1 diabetes, who need to inject themselves frequently with insulin, but tend to do it in the same place. This is a rare medical condition known as injection amyloidosis. Insulin-derived amyloidosis is a rare, yet a significant complication of insulin therapy, because it exacerbates diabetes symptoms.

References

[1] Whidden M, Ho A, Ivanova MI, Schnell S, Competitive reaction mechanisms for the two-step model of protein aggregation. Biophysical Chemistry 193–194: 9–19, 2014.

[2] Ivanova MI, Sievers SA, Sawaya MR, Wall JS, Eisenberg D, Molecular basis for insulin fibril assembly, Proceedings of the National Academy of Sciences USA 106: 18990–18995, 2009.

© Springer Nature Switzerland AG 2020
F. Matthäus et al. (eds.), *The Art of Theoretical Biology*, https://doi.org/10.1007/978-3-030-33471-0_27

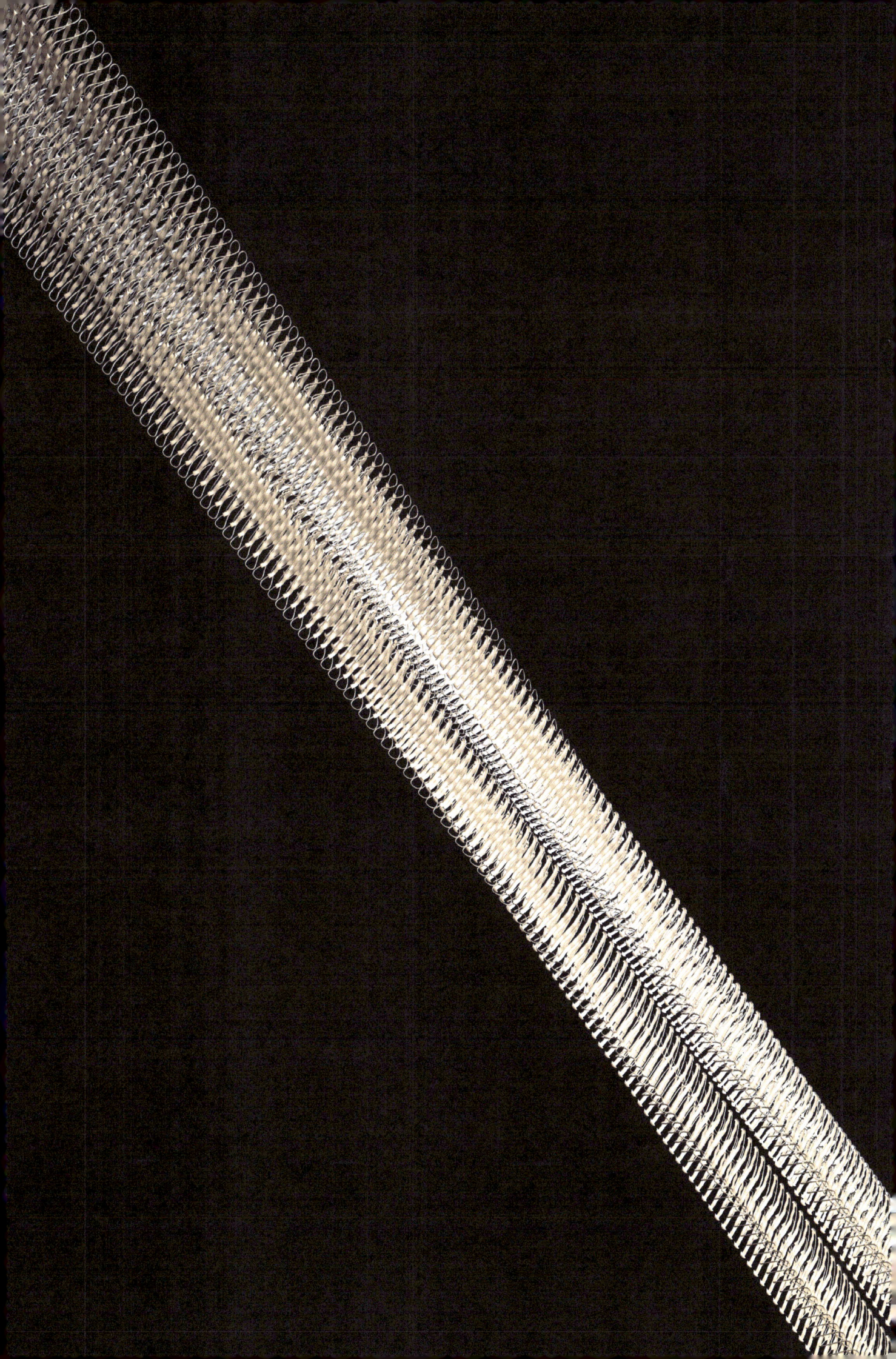

Nothing Stands Still in the Streams of Life

By Alexandre Gouy & Diana Ivette Cruz Dávalos

The research story

This image represents three replicates of a simulation of the evolution of a population of bacteria over time. Each square corresponds to an individual and each color corresponds to a new mutant. The horizontal axis corresponds to the time in generations, and the vertical axis corresponds to the relative abundances of each variant. The fate of this population is driven by various evolutionary forces that lead to constant changes in this population. For example, some mutants can invade the population if they carry an advantageous mutation. Simulating these evolutionary dynamics allows to better understand how evolutionary forces shape observed patterns of diversity or how organisms respond to changes in their environment. A typical application is the study of antibiotic resistance in bacteria.

The image

The choice of colors is central in this image. We put an impressionist touch to this figure by taking them from colors found in impressionist paintings. For each of the tree panels, we loaded a digital image of the painting, and computed the dominant colors of that image. Then, these colors have been randomly assigned to the different variants in our evolving population. The three panels correspond to three different painters and paintings, from the top to the bottom: Auguste Renoir – Bal du Moulin de la Galette (1876); Claude Monet – The Japanese Footbridge (1899); Vincent Van Gogh – The Starry Night (1889).

F. Matthäus et al. (eds.), *The Art of Theoretical Biology*, https://doi.org/10.1007/978-3-030-33471-0_28

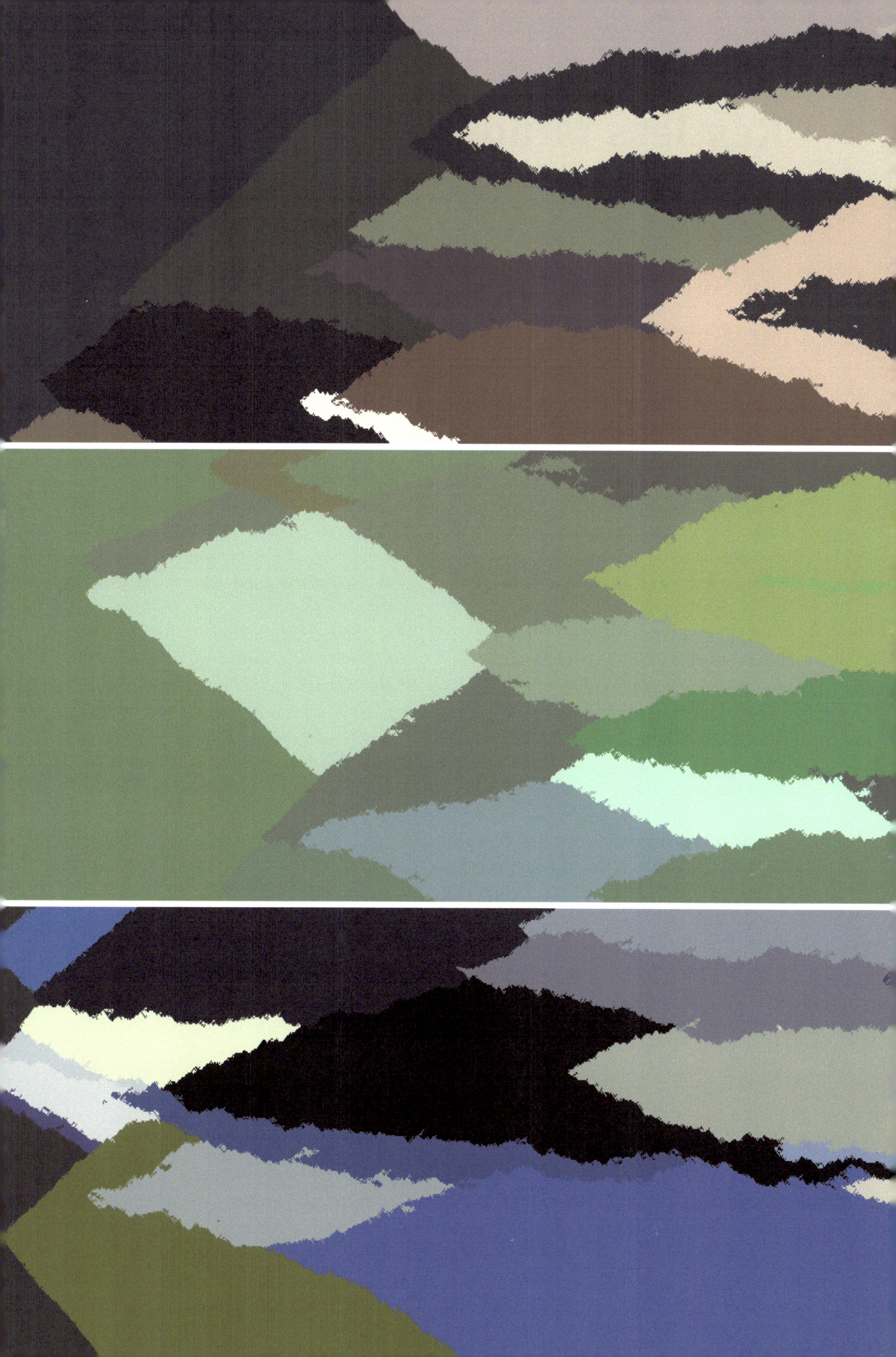

Restless Mind Wandering

By Stefan Fuertinger & Kristina Simonyan

The research story

Epilepsy affects more than 50 million people of all ages and ethnicities worldwide. About one-third of patients develop medication-resistant seizures, for which surgical resection of the brain region causing seizures remains the best treatment option. Surgeons identify these seizure onset zones using so-called intracranial electroencephalographic (iEEG) recordings, which are acquired by opening a patient's skull and implanting electrodes directly on the surface of the brain to record its electrical activity. We used ideas rooted in network science and graph theory to analyze iEEG recordings of epilepsy patients during seizure free periods as well as before, during and after seizure episodes [1]. Specifically, we constructed networks capturing the relation between brain activity as recorded by each individual electrode. Interpreting electrodes as network nodes allowed us to investigate the formation and break-up of nodal communities over the course of a seizure event. We found that epileptic seizures were characterized by a pronounced break-down of nodal community structure well before seizure onset. This strategy allowed us to identify network characteristics of upcoming seizure events.

The image

The depicted computer-generated graph illustrates the formation and break-up of communities in an iEEG network during a ten-minute seizure episode. Every network community is represented using a distinct color. Every time a node migrated from community to another, we rendered an arc whose color indicated the node's target community. The white inset shows a segment of the raw iEEG signal as recorded by an electrode adjacent to the seizure onset zone.

Reference

[1] Fuertinger S, Simonyan K, Sperling MR, Sharan AD, Hamzei-Sichani F, High frequency brain networks undergo modular breakdown during epileptic seizures. Epilepsia 57: 1097–1108, 2016.

© Springer Nature Switzerland AG 2020
F. Matthäus et al. (eds.), *The Art of Theoretical Biology*, https://doi.org/10.1007/978-3-030-33471-0_29

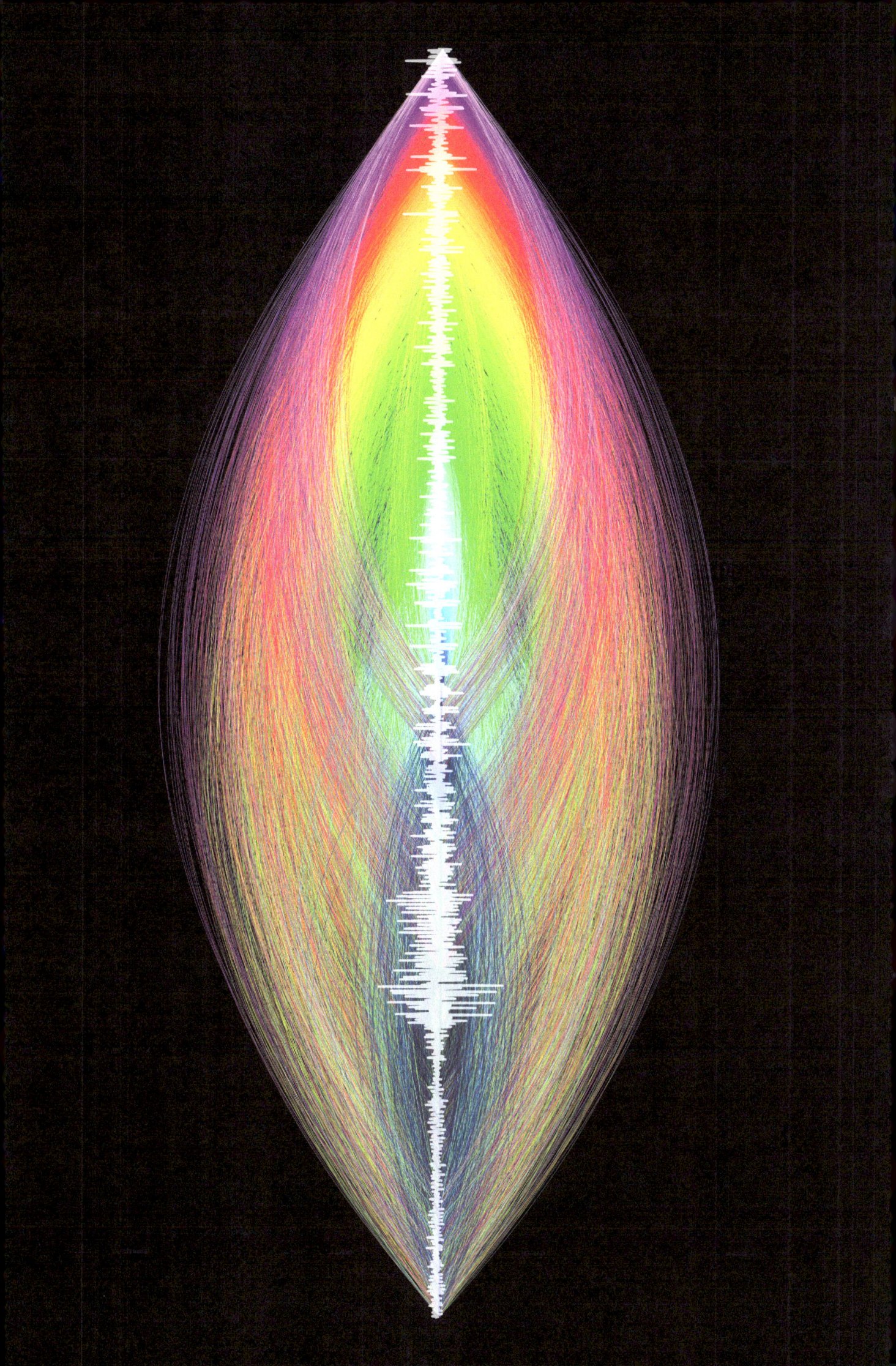

Morphological Echoes

By Mark Robertson-Tessi & Alexander R. A. Anderson

The research story

An important part of tumor progression is the transition from a slow-growing mass to a fast-growing invasive carcinoma. One mechanism that tumor cells use to invade is by acidification of the environment. Tumor cells accomplish this by changing their metabolism to rely more on glucose (sugar) for their energy needs. This leads to an acidic condition that can destroy the normal tissue adjacent to the tumor edge. This erosion provides space for the tumor to grow at a faster rate. In addition to changing their metabolism, tumor cells must also become insensitive to this acidity themselves, so that they can survive.

The image

We simulate the process of a tumor transitioning from benign to aggressive using a mathematical model of evolution. Nine different progressions of this evolution are simulated and colored based on the type of tumor cells present. The tumors grow in a normal tissue (colored gray) that has blood vessels within it (colored white). Each tumor begins as a small clump of benign cells and is allowed to grow and evolve. The image captures the moment at which each tumor starts invading the surrounding tissue. Benign cells are shown in green. Those resistant to the acidosis are colored purple, and the invasive cancer cells, which have heightened glucose consumption, are shown in pink. One can see the emergence of invasive lobes where the pink cells dominate. There is a great variation in structure and shape of the evolved populations.

Reference

[1]   Robertson-Tessi M, Gillies RJ, Gatenby RA, Anderson AR. Impact of metabolic heterogeneity on tumor growth, invasion, and treatment outcomes, Cancer Res 75(8): 1567–79, 2015.

© Springer Nature Switzerland AG 2020
F. Matthäus et al. (eds.), *The Art of Theoretical Biology*, https://doi.org/10.1007/978-3-030-33471-0_30

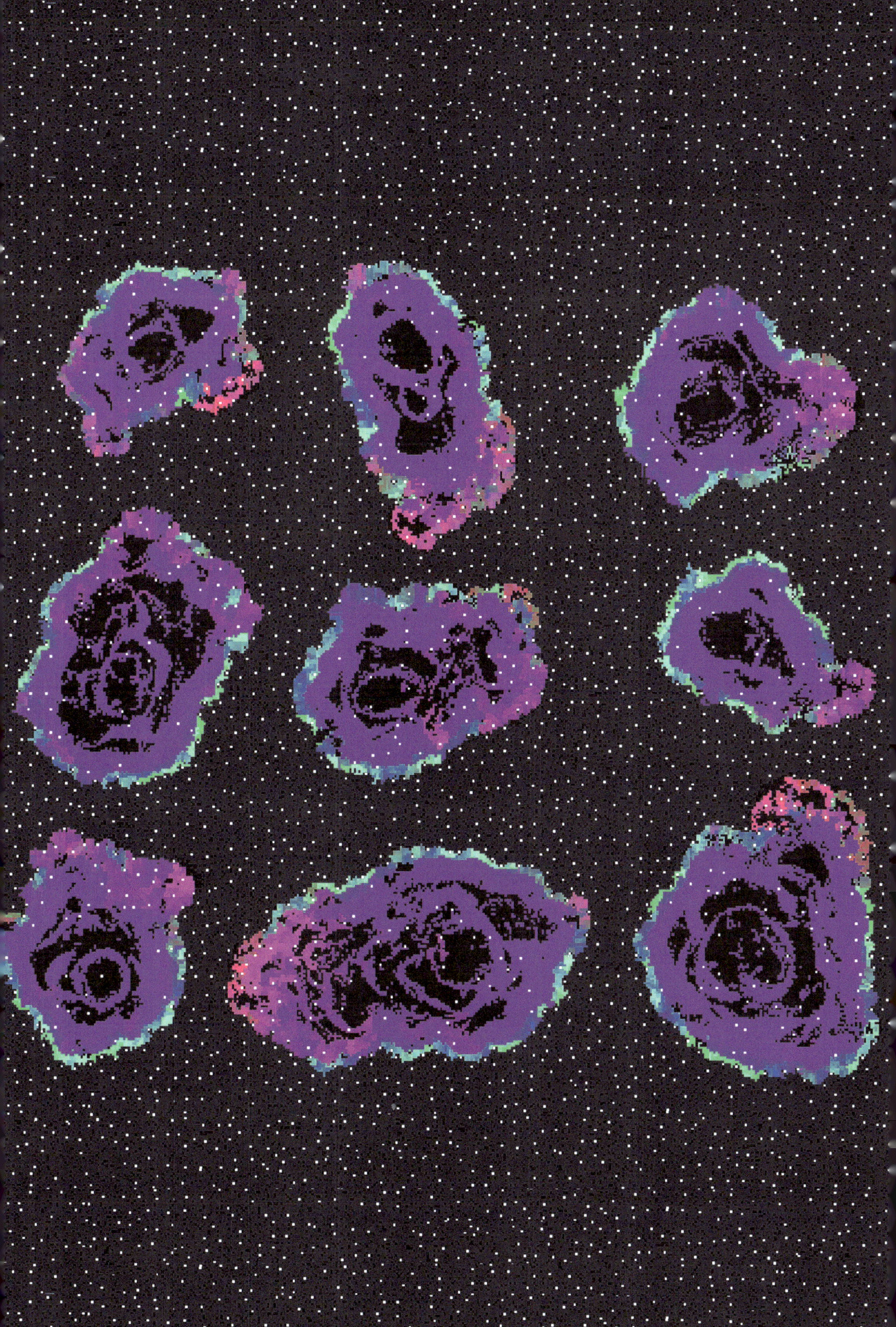

Cancer as a Killer Tsunami

By Jan Poleszczuk, Urszula Foryś, Monika J. Piotrowska & Marek Bodnar

The research story

Back in 2013 our group made an attempt to bring additional insight into the process of carcinogenesis by using the tools of mathematical modeling [1], an approach being at the very core of theoretical biology. We built upon the mathematical model developed by R. Ahangar and X. B. Lin [2] which described carcinogenics as a dynamical and non-linear process of accumulating subsequent mutations. In their model a malignant cell was defined as one that acquired at least n mutations. Our idea for the study was to include a time delay in the model, i.e. make the future state of the process dependent not only on the present state, but also on the events that preceded it, which allowed us to approximate all subsequent n mutation states by a single delay. Moreover, we have decided to consider the spatial aspect by directly modeling cellular motility.

The image

In the image we can see a waves-resembling model-predicted change of malignant cell density in time and space. The model solution presented was obtained numerically using a combination of Mathematica and MATLAB scientific computing software and corresponds to the system parameters reflecting harsh environmental conditions and a large value of time delay (the larger the delay, the further into the past we are looking and the larger the value of approximated number of subsequent mutations n). We can see in this setting that a time delay has a destabilizing effect, i.e. small initial differences in the densities at different points in space are significantly magnified in time.

References

[1] Piotrowska MJ, Foryś U, Bodnar M, Poleszczuk J, A simple model of carcinogenic mutations with time delay and diffusion, Mathematical Biosciences And Engineering 10(3): 861–872, 2013.

[2] Ahangar R, Lin XB, Multistage evolutionary model for carcinogenesis mutations, Electron. J. Diff. Eqns. 10: 33–53, 2003.

© Springer Nature Switzerland AG 2020
F. Matthäus et al. (eds.), *The Art of Theoretical Biology*, https://doi.org/10.1007/978-3-030-33471-0_31

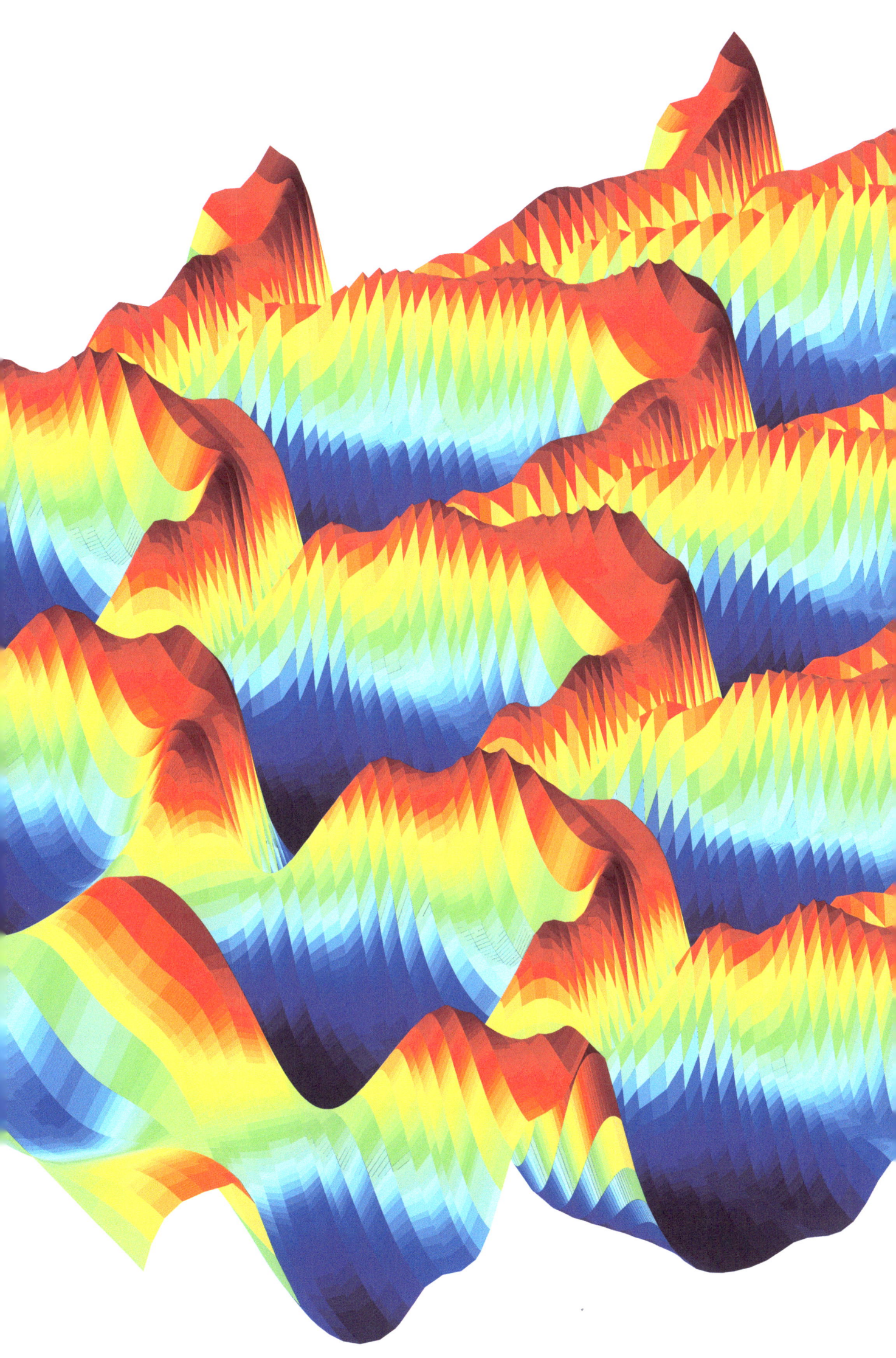

Cells Are Watching You

By Gaëlle Letort

The research story

The development of organs is a very complex system, involving numerous processes that have to be precisely coordinated. In particular, during the formation of the olfactory system in the zebrafish embryo, stem cells migrate toward specific places on either side of the brain, and form a compact group of cells called placodes. At the same time, those cells differentiate into olfactory sensory neurons. To understand the steps controlling olfactory placode formation, biologists study cell behavior in real time during this process and the genes involved in neuron specification (Dr. Patrick Blader laboratory, CBI, Toulouse). In particular, they investigate how molecules involved in cell migration or guidance cues, when mutated, will affect the overall cell behavior and placode formation. To help to interpret the effect of different mutants during cell migration, we developed a mathematical model of cell motion according to specific cues.

The image

This image is the result of a simulation of zebrafish placode formation in the absence of extra-cellular matrix, and a chemotaxis source present at the center. It shows the trajectory of the simulated cells (different color lines) and their final position in white. Simulations were performed and the plot was generated with Matlab. Colors were inverted (hence the black background) and light was added in the "eye" center and luminosity was arranged with Gimp for visual effect.

© Springer Nature Switzerland AG 2020
F. Matthäus et al. (eds.), *The Art of Theoretical Biology*, https://doi.org/10.1007/978-3-030-33471-0_32

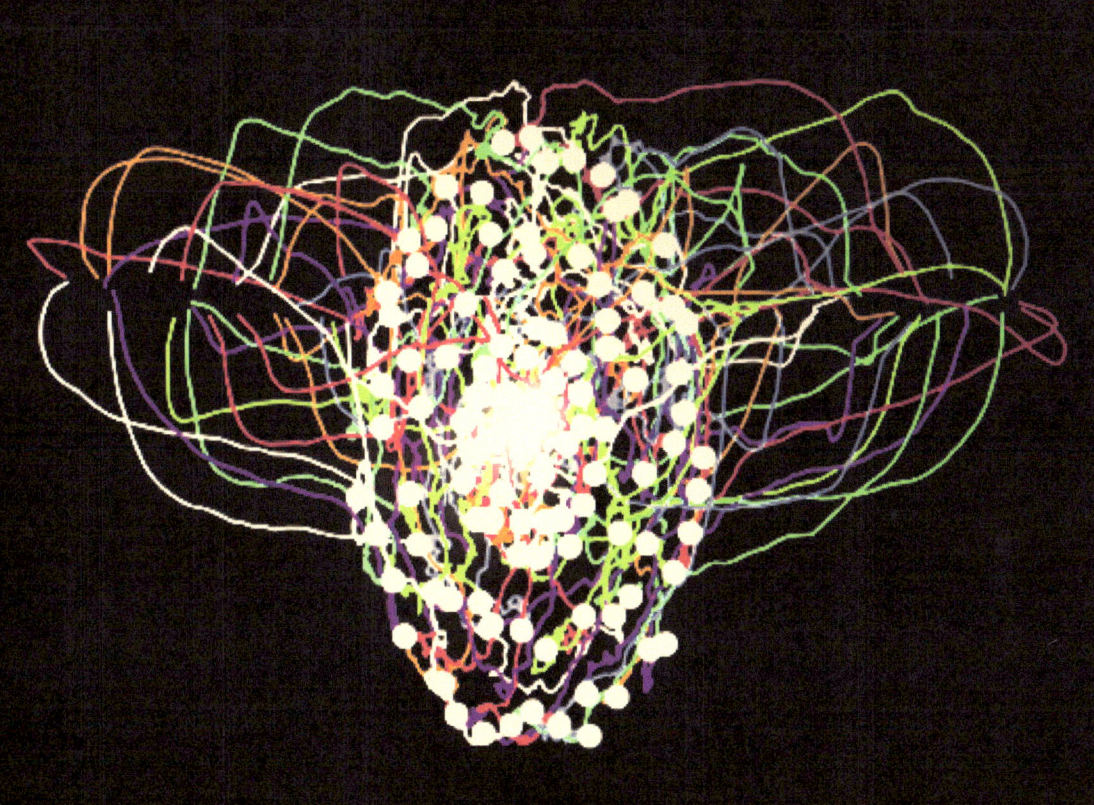

Roots or Flowers? Take a Guess…

By Ishan Ajmera, Leah Band, Dov Stekel & Charlie Hodgman

The research story

Phosphorus is a growth-limiting nutrient for plants. The use of phosphate (Pi) fertiliser is unsustainable because its stocks are non-renewable and the run-off of surplus fertiliser damages the environment. It is therefore desirable to develop crops with increased Pi uptake efficiency by their roots. The latter consist of layers of different cell types and fluid-filled intercellular gaps. In response to Pi deficiency, certain root cells die off forming air spaces, called aerenchyma, which have narrow fluid-filled membrane-bound gaps crossing them. We are investigating the impact of this change in roots using a multi-cell modelling framework, OpenAlea.

The image

The images depict the simulated distribution of Pi across an actual cross section of a rice root. The colours represent cellular Pi levels: red is highest and dark blue is zero. The top image shows the general distribution, while the central image shows the effect of manually introducing aerenchyma. It appears that the fewer cells reduce the barriers to inward flow of Pi through the fluid-filled gaps, thus increasing its uptake and lowering Pi levels in the outer cells. The third image shows the situation when flux through intercellular and fluid-filled gaps are blocked. The Pi distribution becomes inverted, with higher levels external to the aerenchyma, because it can no longer cross that space, so the Pi is drawn from the other root cells at a higher rate. Altogether, this suggests that the presence of aerenchyma increases Pi uptake in roots and plant breeders should aim to develop crop varieties with a higher proportion of aerenchyma.

F. Matthäus et al. (eds.), *The Art of Theoretical Biology*, https://doi.org/10.1007/978-3-030-33471-0_33

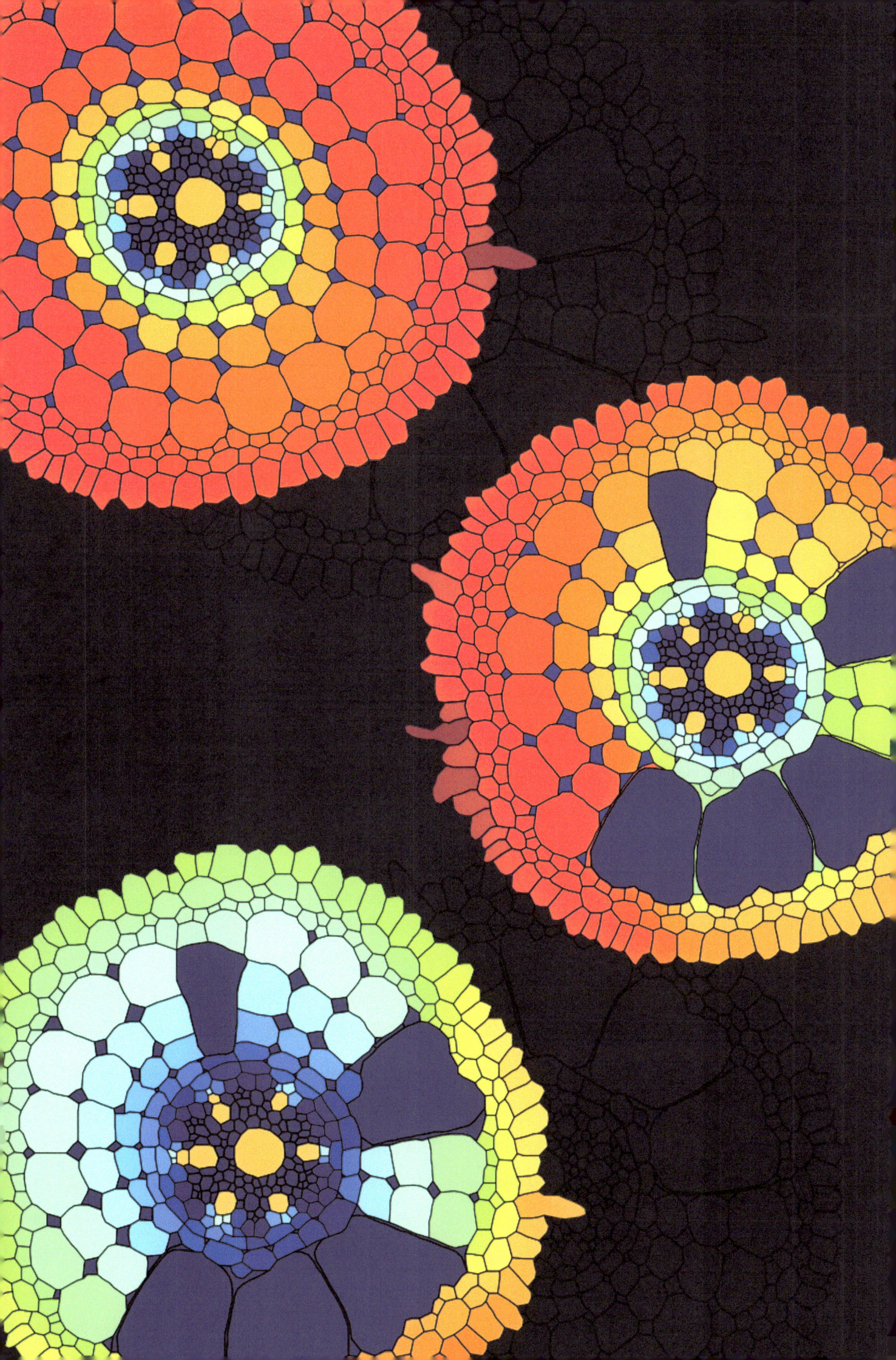

Spectral Forms and Cosmic Storms

By Michael Colman

The research story

Cycling of calcium ions is a critical component of the link between the electrical activity and mechanical contraction of heart-muscle cells, and depends on complex, microscopic structures within each cell. The distributed nature of these structures can promote the emergence of an abnormal and potentially dangerous phenomenon: a spontaneous calcium wave, initiated and maintained by random oscillations in the proteins responsible for electro-mechanical coupling. This image shows a snapshot of calcium concentration in the 3D volume of a cell during a calcium wave (blue-yellow-pink colours) simulated using a computational model implementing an idealised cellular structure; the complexity emerging even under these idealised conditions was striking.

The image

The visualisation approach aimed to invoke the enhanced images of cosmic dust and stellar nurseries found in the vast collection of stunning astronomy photos; what was unintended is the impression of various ethereal faces and forms which seem to slowly emerge as one's eyes relax on the image, much like the "magic eyes" of old. This author personally delights in the representation of multiple scientific disciplines within this image: a biological process was captured in a computer model using the approximations of physics implementing mathematical methods; the visualisation was inspired by cosmology and highlights the innate psychological processing involved in human facial and pattern recognition.

Reference

[1] Colman MA, Pinali C, Trafford AW, Zhang H, Kitmitto A, A computational model of spatio-temporal cardiac intracellular calcium handling with realistic structure and spatial flux distribution from sarcoplasmic reticulum and t-tubule reconstructions, PLoS Comput Biol. 13(8): e1005714, 2017.

© Springer Nature Switzerland AG 2020
F. Matthäus et al. (eds.), *The Art of Theoretical Biology*, https://doi.org/10.1007/978-3-030-33471-0_34

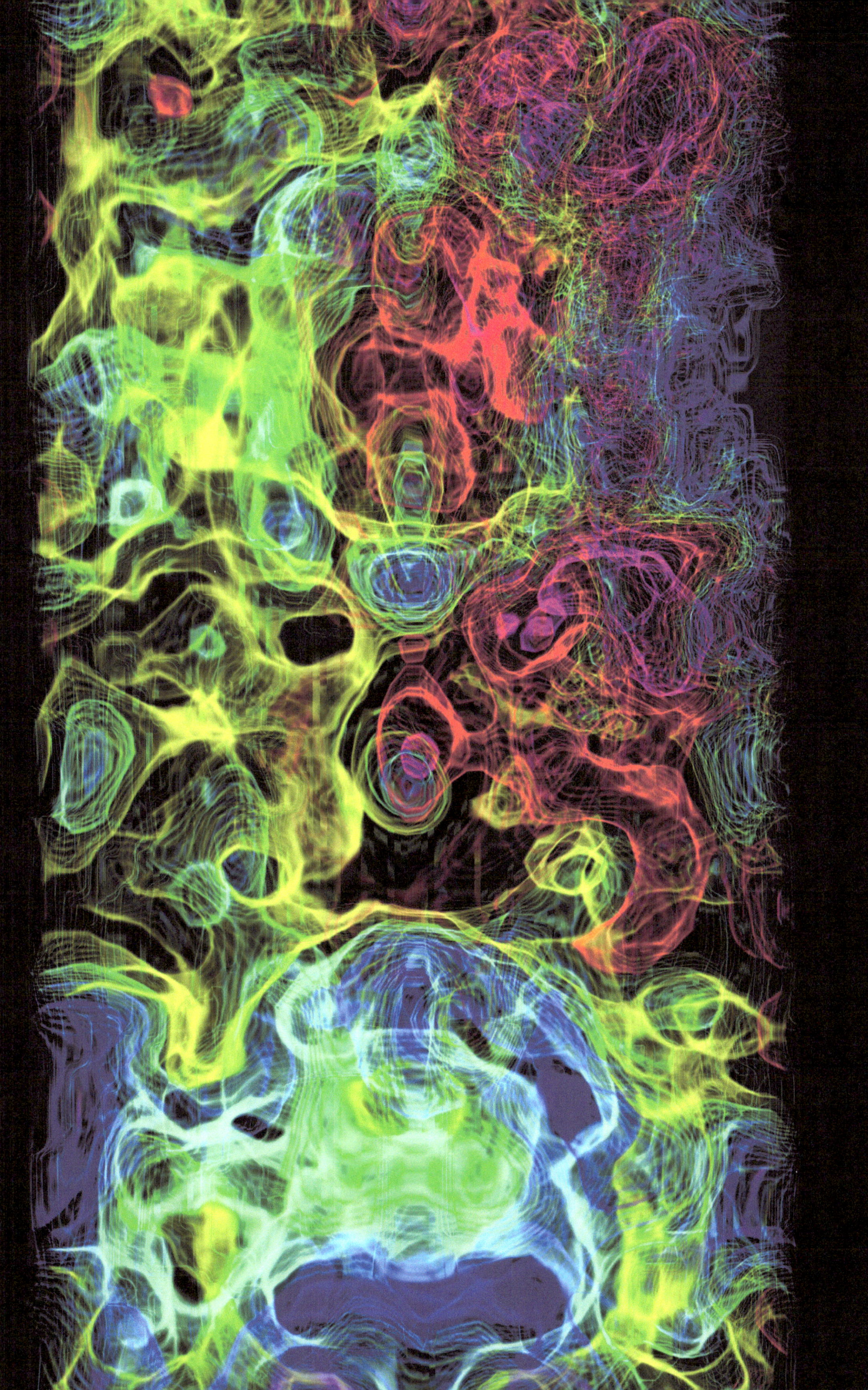

Antigenic Explosion

By Chandler Gatenbee & Alexander R. A. Anderson

The research story

Cancer is driven in part by repeated genetic mutations, generating tumors composed of many distinct clones. Each clonal population expresses different antigens, which are small molecules that can be recognized by the body's immune system. To avoid detection and destruction by immune cells, tumors often go through a process known as immunoediting. Here, we simulate this process to understand how the clonal diversity of a tumor changes with time. This has implications for immunotherapy, a promising new class of treatments that uses the patient's immune system to attack the tumor.

The image

The image shows the results from a simulation in which a tumor successfully mitigated immune attack through immunoediting. Shown is the clonal network of the tumor, where each small circle represents a tumor clone, and lines indicate ancestry, connecting children to their parental clone. Colors represent the type of tissue to which each clone belongs: blue are from normal tissue; green exist within a benign adenoma; red compose a malignant carcinoma. In this case, the immune system was able to remove most tumor clones in the normal and adenoma tissues. However, the carcinoma eventually developed an immunoediting strategy and clones were no longer subject to elimination by the immune system, despite expressing antigens that would normally elicit a response. An explosion of immunogenic antigens can therefore be used to infer a high degree of immunoediting in a patient's tumor. This can inform decisions about whether or not a patient is a good candidate for different types of immunotherapies.

© Springer Nature Switzerland AG 2020
F. Matthäus et al. (eds.), *The Art of Theoretical Biology*, https://doi.org/10.1007/978-3-030-33471-0_35

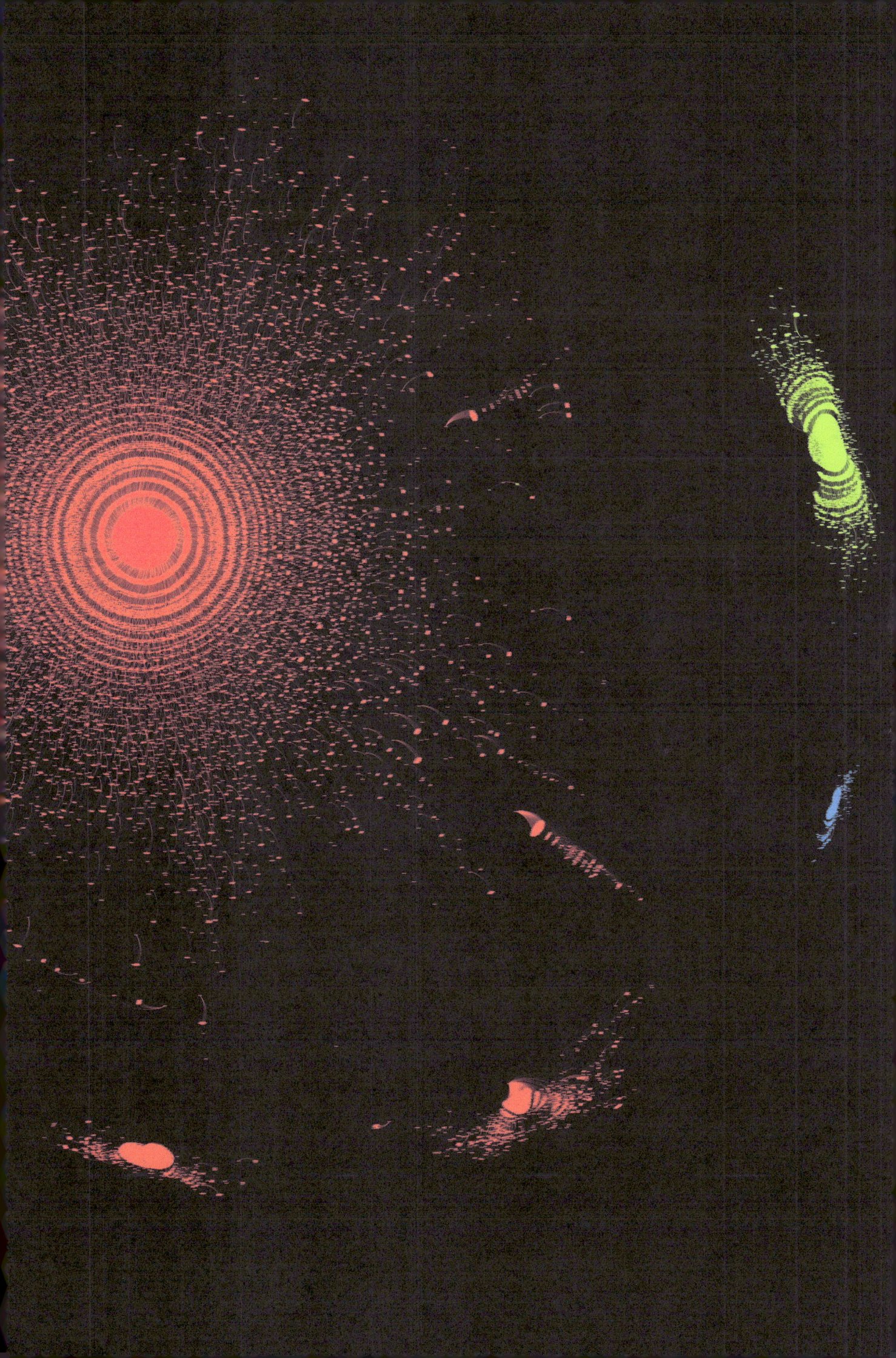

Crop Circles of Cancer

By Katharina Baum, Jagath C. Rajapakse & Francisco Azuaje

The research story

Despite research advances, cancer is still often a deadly disease. Over the last decade, the research community has gathered abundant and diverse measurements of multiple layers of molecular data in thousands of tumors. Considering the measured molecules in their network context, i.e. how genes, proteins or metabolites interact, is key to a mechanistic understanding of why some patients respond well to a drug, whereas for other patients therapy fails. We performed network-based analyses to detect molecular differences between patient groups, and to derive novel drug targets from this data. An important step in our analyses is to take into account the hierarchical, modular organization of the underlying biological systems of tumors. This can improve our ability to detect dysregulations which point to more effective, personalized therapeutic options.

The image

The picture shows mRNA-mRNA and protein-protein interaction networks derived from tumor data of breast cancer patients (from TCGA, CPTAC). The interactions are established by correlating such measurements, which were subsequently truncated at two different association strengths. The hierarchy of the biological networks is captured by fitting them to "stochastic block models", which allow the detection of hierarchical structures as shown in the figure [1]. Each colored square is one block at the lowest hierarchy level, that is a collection of proteins or mRNA molecules with similar network characteristics. The size of the squares represents the number of assigned molecules, the thickness of the arcs scales with the number of interactions between the blocks. Fit and graphical representation have been performed with graph-tool [2].

References

[1] Baum K, Rajapakse JC, Azuaje F, Analysis of correlation-based biomolecular networks from different omics data by fitting stochastic block models, F1000Research 8:465, 2019.

[2] Peixoto TP, The graph-tool python library, figshare, 2014.

© Springer Nature Switzerland AG 2020
F. Matthäus et al. (eds.), *The Art of Theoretical Biology*, https://doi.org/10.1007/978-3-030-33471-0_36

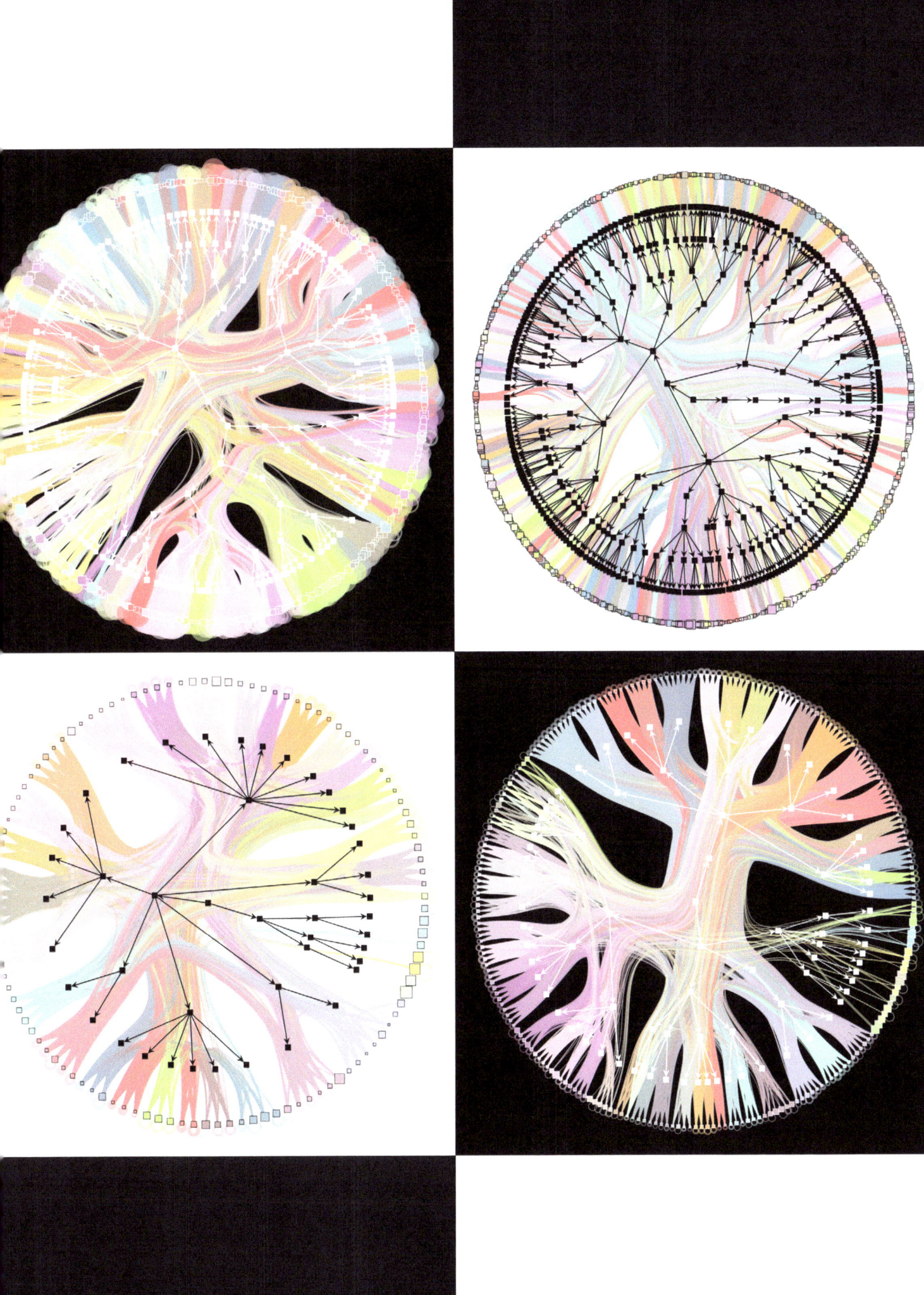

Scalp

By Valerii Sukhorukov & Michael Meyer-Hermann

The research story

The cell is not only the only structure building up living tissue. Every cell contains its own energy factory, termed mitochondria. They build up a network inside the cell, they divide into fragments or merge to form bigger clusters, and they move under the guidance of the cell cytoskeleton. Using computers it is possible to simulate how the mitochondrial network is forming and dynamically changing. Mitochondria contain their own DNA and are, thus, vulnerable to mutations, just like the DNA in the cell nucleus. Our main driving question is to understand the process of mitochondria quality control and damage clearance, which is a process increasingly lost in the course of ageing. We think that the continuous fusion and fission of mitochondria is a critical element of this maintenance process and is particularly effective during cell division [1].

The image

The image shows a computer simulation of a cell on a Petri dish with a nucleus (yellow), a cytoskeleton (red), and a mitochondria network (tan). The cytoskeleton network serves as guidance for mitochondria fragments. Both the cytoskeleton and the mitochondria network are subject to dynamic reorganisation, which is the focus of the computer simulation. The figure shows a snapshot of the computer simulation of the network dynamics [2].

References

[1] Sukhorukov et al. Emergence of the mitochondrial reticulum from fission and fusion dynamics.
PLoS Comput Biol 8(10): e1002745, 2012.

[2] Sukhorukov V, Meyer-Hermann M, Structural heterogeneity of mitochondria induced by the microtubule cytoskeleton.
Sci Rep 5: 13924, 2015.

© Springer Nature Switzerland AG 2020
F. Matthäus et al. (eds.), *The Art of Theoretical Biology*, https://doi.org/10.1007/978-3-030-33471-0_37

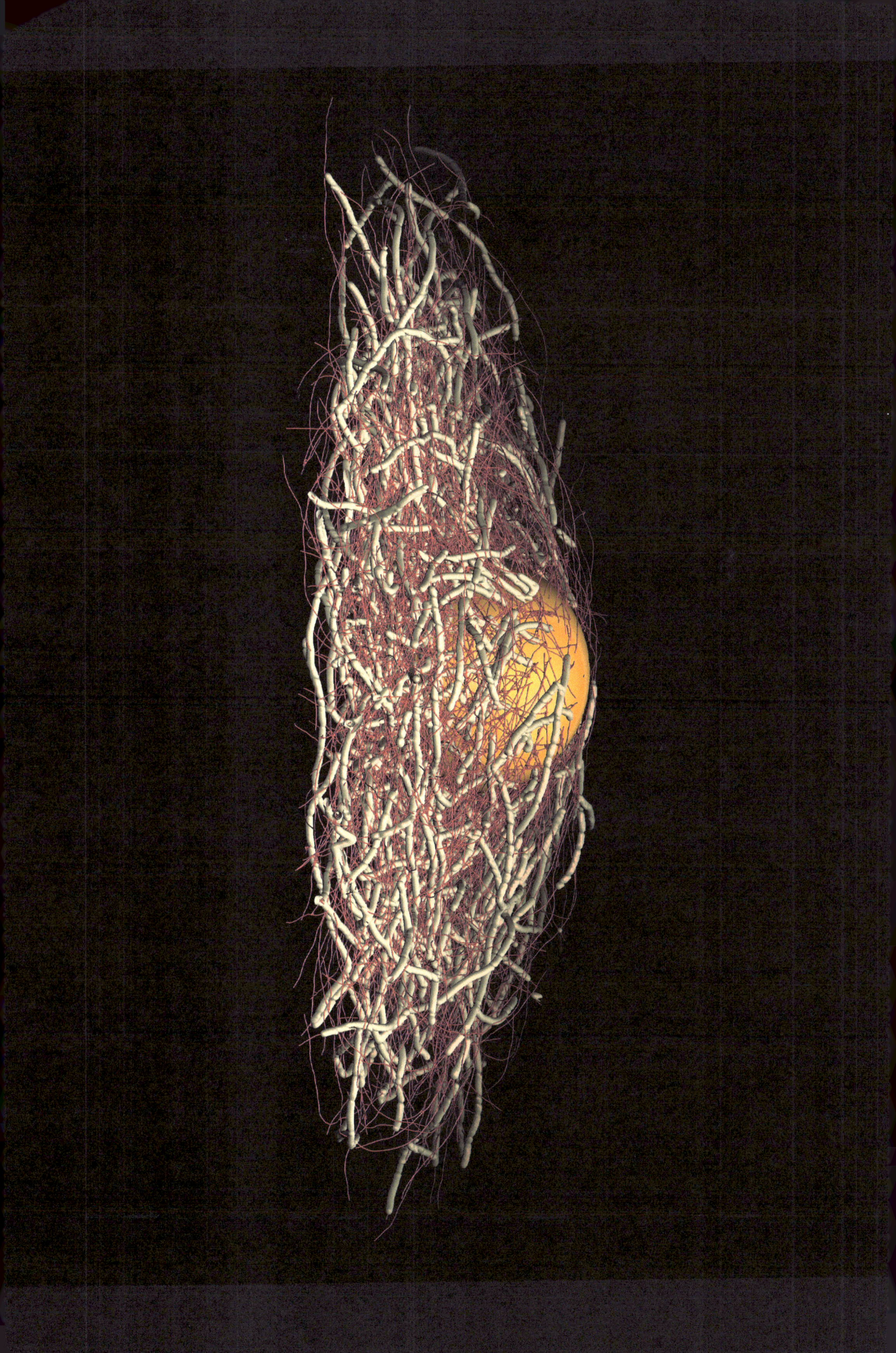

Coupled Invasion

By Jill Gallaher & Alexander R. A. Anderson

The research story

Glioblastoma (GBM) is the most common and aggressive form of brain cancer in both children and adults. Cells in GBM can invade deep into the brain tissue, making complete removal impossible, and chemotherapy after surgery often only kills some of the cells. We are trying to understand how different migration patterns and drug sensitivities of single cells ultimately affect a tumor's response to treatment. This image represents how we weave together different tools to understand this complex problem by showing a mashup of biological and computational data that we combine to quantify a spatial map of tumor progression. The white matter region of the brain tissue can be distinguished in the image along the diagonal from the top right, where the tumor initiated, to the middle left, at the tumor edge, where cells are observed to move faster.

The image

Red and green fluorescent images from two different cell types invading into a rat brain were enhanced for contrast and masked to occupy the upper left half of the tumor. The corresponding computational model image was made using Java library graphics, which was then contrast and edge enhanced and masked to occupy the lower right half of the tumor. The interface between these images was manipulated to create cohesion. Single cell trajectories recorded over the course of the experiment are shown as black lines, which were plotted in MATLAB and laid over, while a white-gray background gradient was laid under. Pixelmator software was used for all image manipulations.

© Springer Nature Switzerland AG 2020
F. Matthäus et al. (eds.), *The Art of Theoretical Biology*, https://doi.org/10.1007/978-3-030-33471-0_38

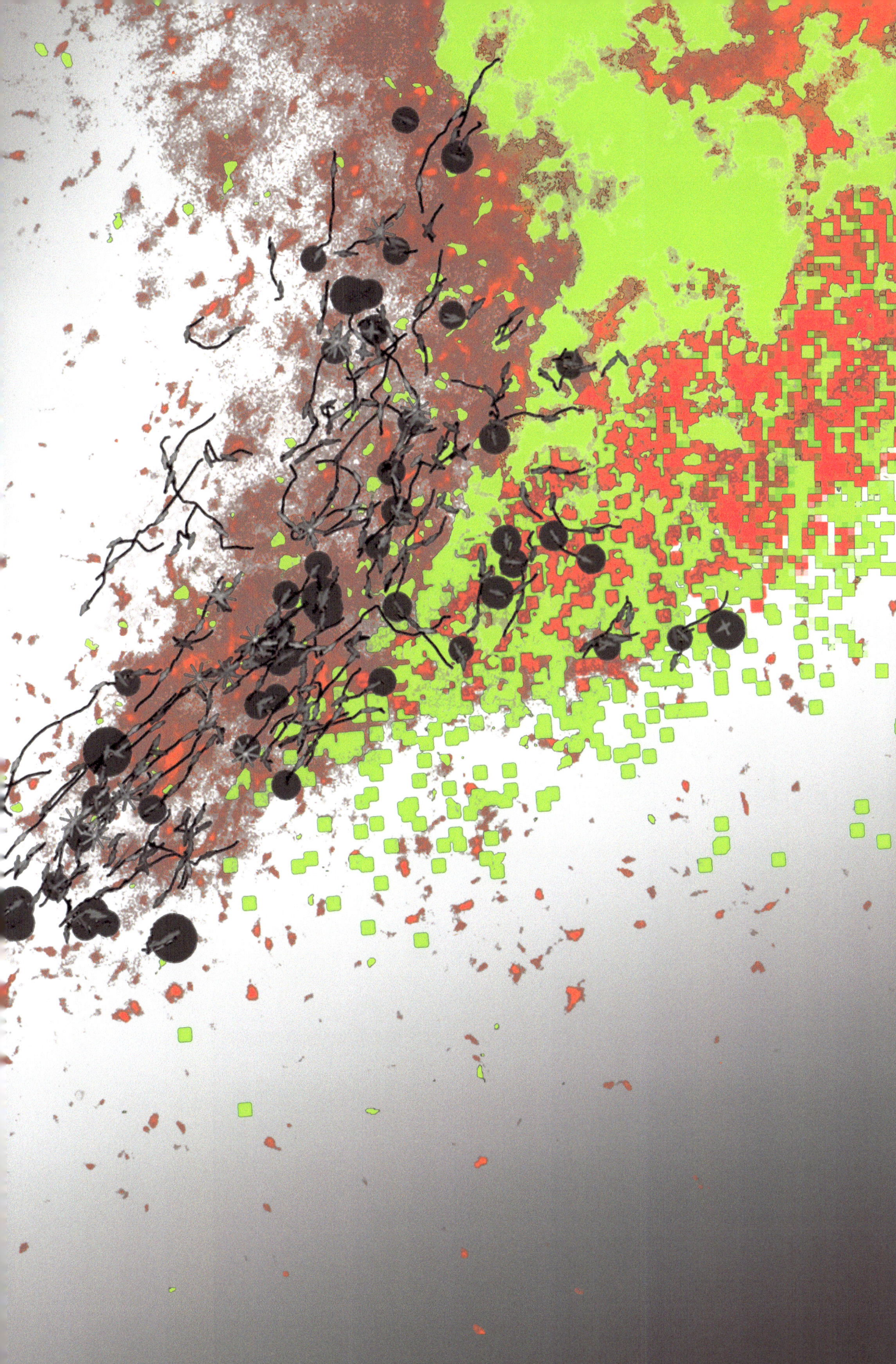

Lost in the Cells

By Nina Kudryashova & Alexander Panfilov

The research story

The heart consists of several cell types. With ageing, the main cells (cardiomyocytes) are gradually replaced by the cells of connective tissue (fibroblasts), which keep the tissue together but perturb electrical signals required for the synchronous functioning of the cardiac cells. Fibroblasts and cardiomyocytes form different patterns. However, the mechanism behind this patterning was not clear. Here, we analyze the pattern formed in a petri dish in laboratory experiments.

The image

It turned out that the cardiac cells (black) construct a MAZE to remain connected. It is not very easy to find a way through it, right? However, we had to solve this problem to find out if a particular network of cardiac cells is interconnected. In some way, we can view the pattern of cardiac cells as a system of waterways and electrical signal as a ship navigating through these waters. There are lands of fibroblasts in-between. We coloured each land differently and, as a result, got a useful map for navigation. The shortest way through this water network always lays between the lands of different colours. It is rather easy to navigate with such a map and avoid taking dead ends leading inland. Also, the characteristic size of these lands told us about the density of this waterway network: the denser it is and the smaller are the lands, the less the ship (the electrical signal) has to go around, and faster it can get from any given point A to point B.

© Springer Nature Switzerland AG 2020
F. Matthäus et al. (eds.), *The Art of Theoretical Biology*, https://doi.org/10.1007/978-3-030-33471-0_39

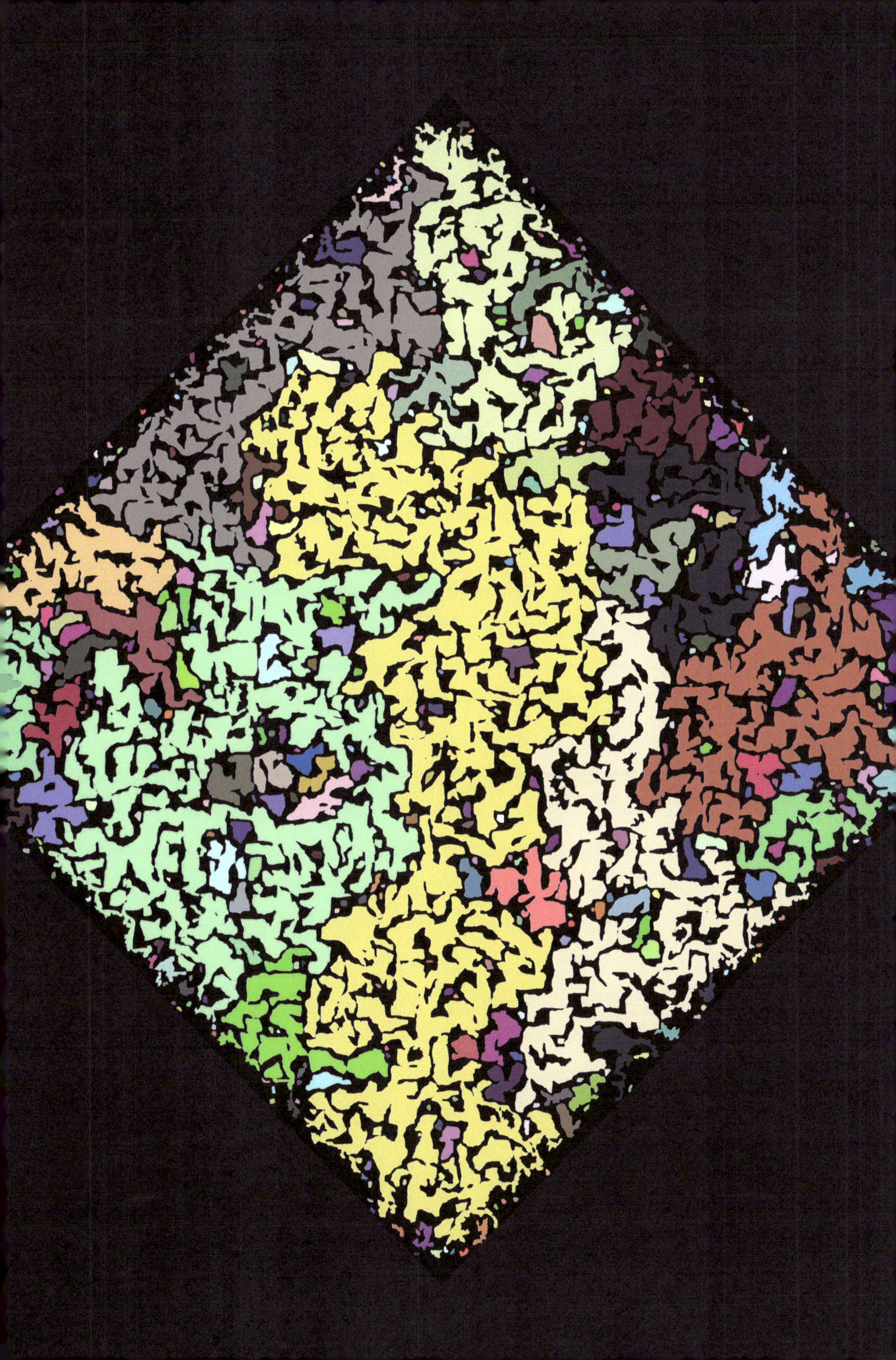

Becoming Important

By Glenn Webb

The research story

A mathematical model for the production of blood cells in the human body uses cell maturity to track the morphological development of blood cells in the bone marrow. This development begins with the most primitive stem cells, and through successive generations of cell division, cells pass through many increasing stages of maturity. Ultimately, the most mature cells become differentiated to specific cell types and enter the circulating blood. The most primitive blood stem cells, which constitute a very small fraction of the total population of all blood cells, are extremely important in the stabilization of the blood cell population. If there is an insufficient supply of these most primitive cells, the blood cell population may exhibit behavior corresponding to diseases such as aplastic anemia and acute leukemia.

The image

The solutions of the differential equations describing the production of blood cells are numerically simulated in the figure. The most immature stem cells have a very small number at any time. As time advances these cells divide and proliferate to increasingly mature cells, with increasingly large numbers. The gray panels show the maturity distribution of cells at different time points, with the most mature cells at the highest level of maturity values.

References

[1] Bernard S, Pujo-Menjouet L, Mackey MC, Analysis of cell kinetics using a cell division marker: mathematical modeling of experimental data, Biophys. J., Vol. 84: 3414–3424, 2003.

[2] Dyson J, Villella-Bressan R, Webb GF, A Nonlinear age and maturity structured model of population dynamics, J. Math. Anal. App. 242: 93–104, 2000.

© Springer Nature Switzerland AG 2020
F. Matthäus et al. (eds.), *The Art of Theoretical Biology*, https://doi.org/10.1007/978-3-030-33471-0_40

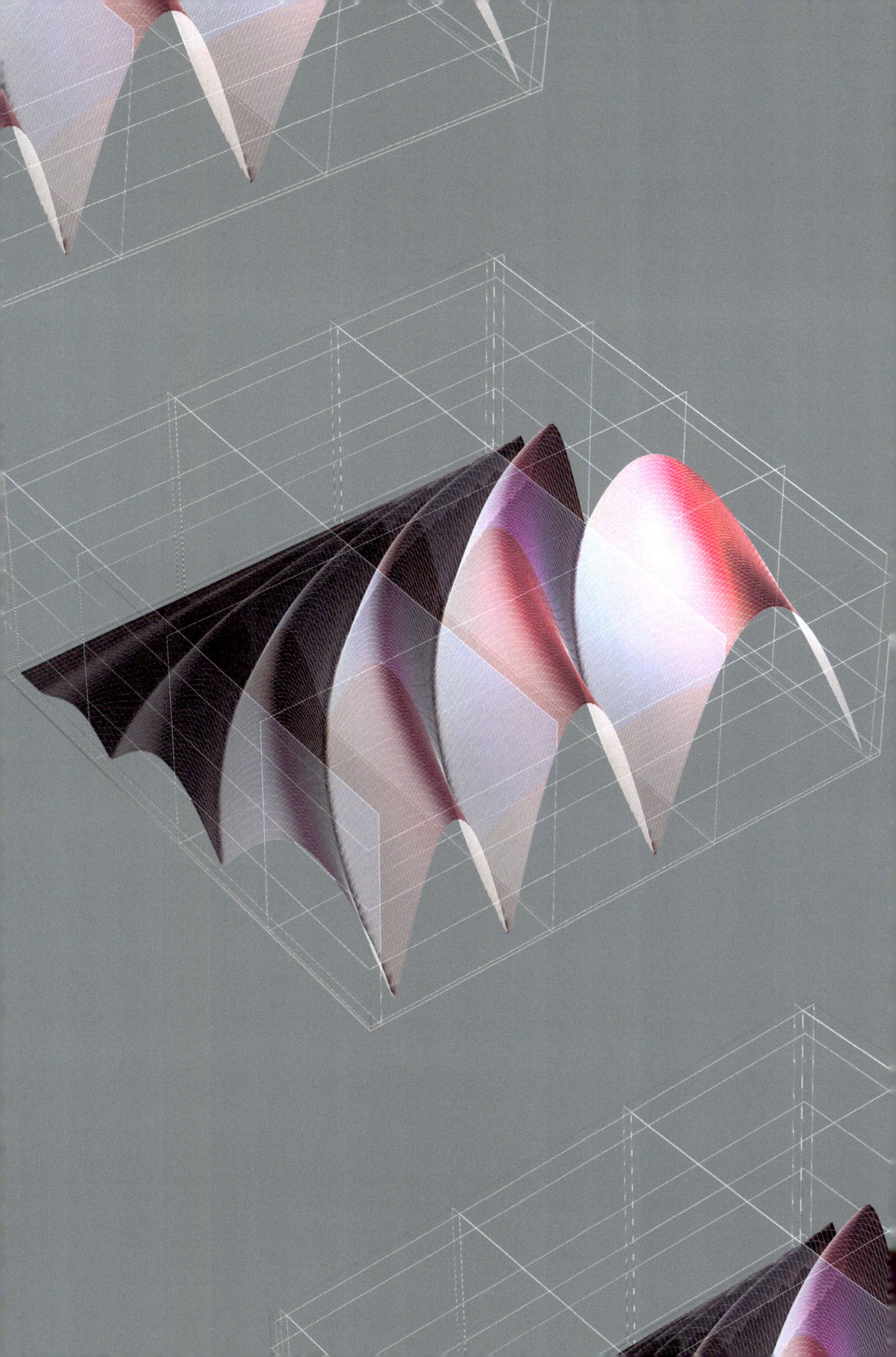

Community Matters

By Anna Lewis, Nick Jones, Mason A. Porter & Charlotte Deane

The research story

Understanding biochemical networks is a focus of modern biological sciences. These networks describe how proteins interact with other proteins and with genes, how genes are expressed, and how biological information is processed. There can be thousands of proteins involved. One way to analyze these complex networks is to divide them into "protein communities". Like communities in human societies, these communities contain members that interact closely with each other. And like communities in human societies, the members often have a shared activity. For example, members of a hockey club all play hockey. So we expect that by identifying protein communities we can gain understanding of the functions the proteins are involved in. These communities of proteins exist at multiple scales in a network, with smaller communities often forming parts of larger ones.

The image

In the image we classify proteins into protein communities at varying resolutions. On the very left each protein forms its own community, which is indicated in black. As we increase the community size, we get the colored bands that expand from left to right. At the right end of the figure we end up with one dominating community in blue, which includes all the proteins in the network. Having chosen to color the proteins that stand alone black, we create a pleasing effect that highlights the colorful protein communities.

Reference

[1] Lewis ACF, Jones NS, Porter MA, Deane CM, The function of communities in protein interaction networks at multiple scales, BMC Systems Biology 4: 100, 2010.

© Springer Nature Switzerland AG 2020
F. Matthäus et al. (eds.), *The Art of Theoretical Biology*, https://doi.org/10.1007/978-3-030-33471-0_41

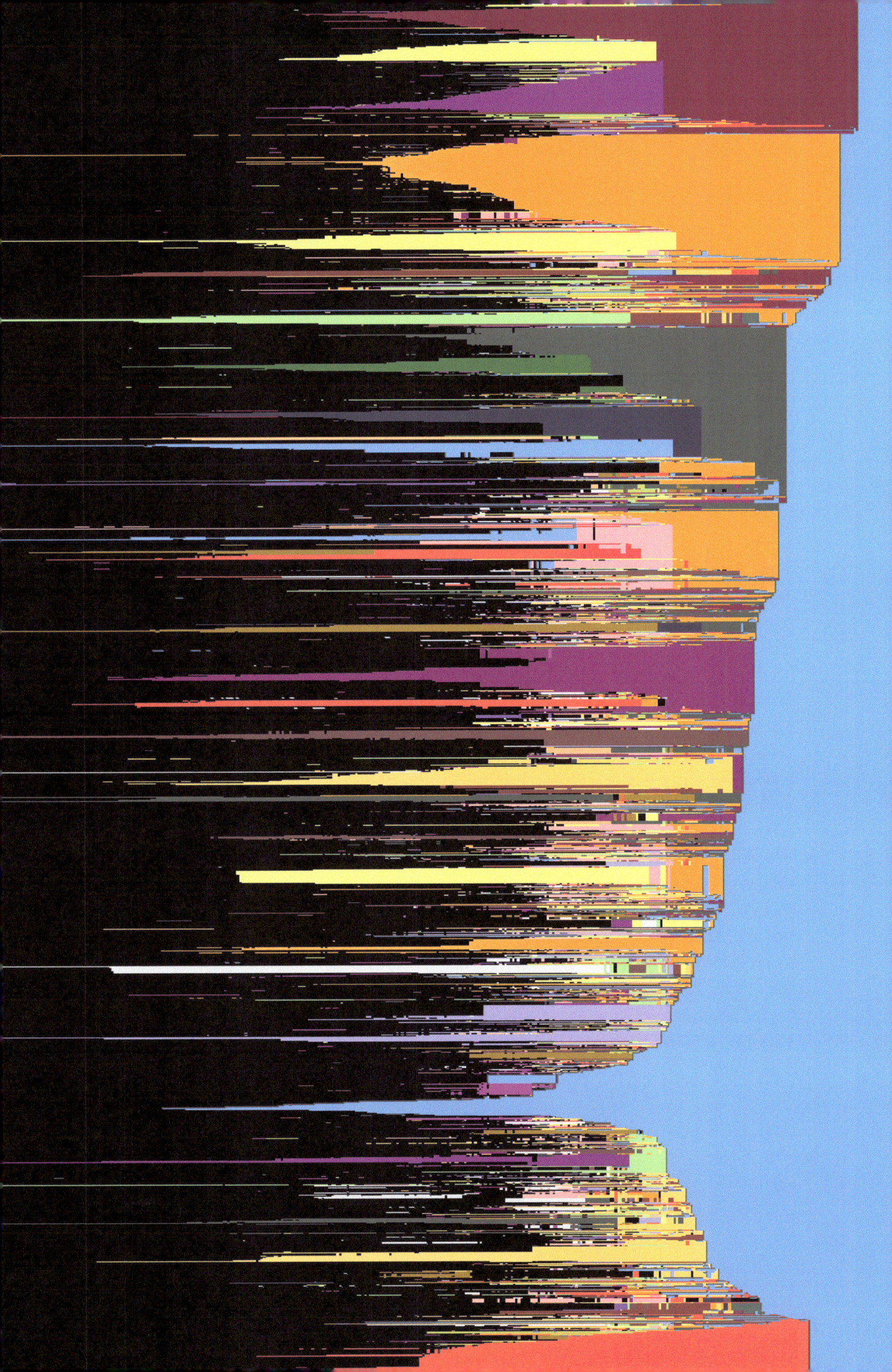

Acidic Dance

By Sandesh Hiremath, Stefanie Sonner, Christina Surulescu & Anna Zhigun

The research story

One of the hallmarks of cancer is the upregulation of glycolysis, both in aerobic and hypoxic conditions, triggering the acidification of the extracellular region, while normal cells have a reduced capability of surviving at low pH values. This seems to confer tumor cells several advantages, among others enhancing migration and reducing sensitivity towards chemo- and radiotherapy. Motivated by the classification of histological tumor infiltrative patterns associated to oncological outcome, in [1] we developed and studied a mathematical model for acid-mediated tumor invasion. The equations couple a chemotaxis system for the motion and spread of tumor cells biased by the gradient of extracellular pH with a stochastic differential equation describing the dynamics of intracellular acidity. Being inherent to most biological processes, randomness is a relevant feature, also on the level of individual cells. In particular, the exchange of protons across the cell membrane has the role of balancing the intra- and extracellular acidity, and is highly erratic, as experiments show. Our model characterising the acid-mediated space-time evolution of cancer cell population density is able to qualitatively reproduce all known types of infiltrative patterns.

The image

The picture shows a sequence of simulated tumor and acidity patterns (contour lines and color patches, respectively) at three successive times. Warmer/lighter colors: high densities of cells and concentrations of acid, colder/darker colors: low densities and concentrations. The most acidic regions correspond to large cancer cell aggregates.

Reference

[1]   Hiremath S, Sonner S, Surulescu C, Zhigun A: On a coupled SDE-PDE system modeling acid-mediated tumor invasion, DCDS B 22: 1–31, 2017.

© Springer Nature Switzerland AG 2020
F. Matthäus et al. (eds.), *The Art of Theoretical Biology*, https://doi.org/10.1007/978-3-030-33471-0_42

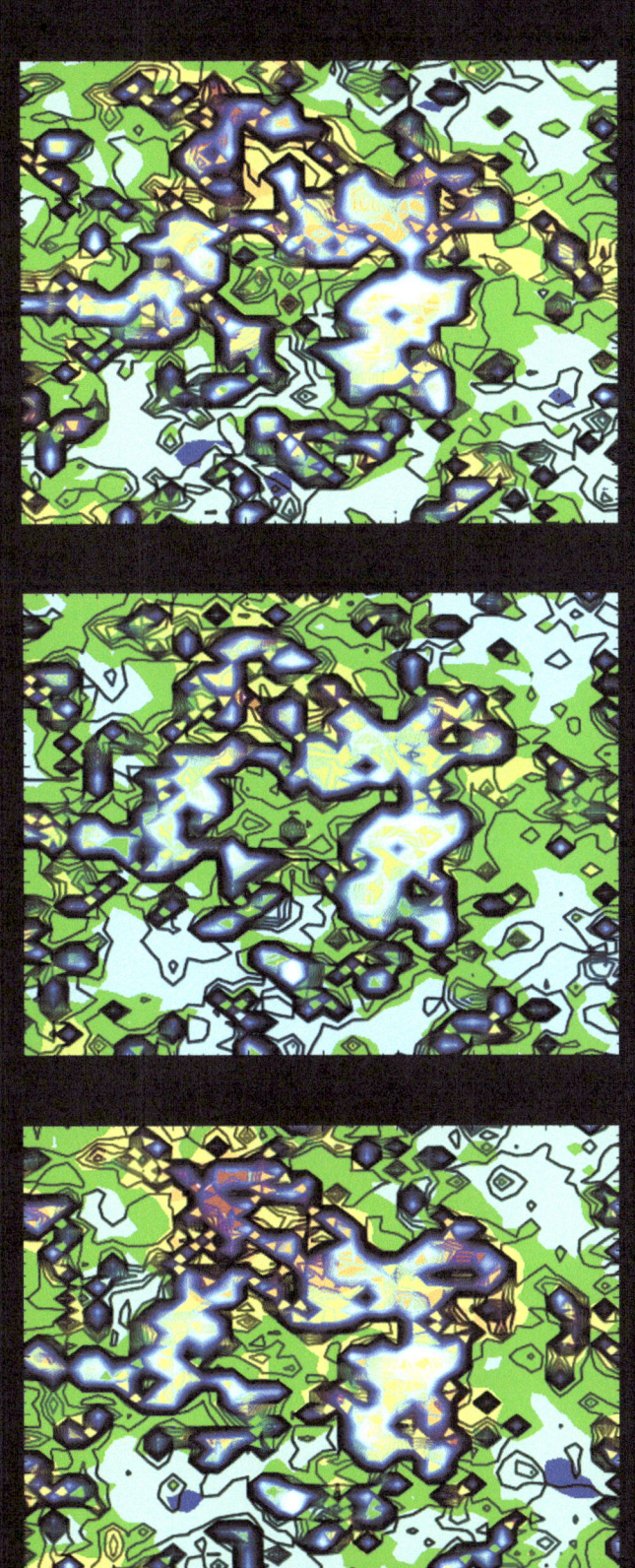

Flocking, Swirling and Spinning Stars in a Cell

By Neha Khetan & Chaitanya Athale

The research story

Microtubule (MT) filaments nucleated from centrosomes form star-like structures or asters. In eukaryotic cells, they act as the 'railway network' of the cell. Cargos ranging in size from mitochondria to smaller vesicles containing neurotransmitters, nutrients or pathogens are all transported on this network. The 'railway' connects the boundary of the cell to it's interior core where the cell nucleus resides. The 'engines' on the 'railway' are molecular motors, consuming metabolic energy and walking on the tracks. However, contrary to the picture of a static 'railroad', the network of astral MTs are constantly undergoing a restructuring due to intrinsic changes in lengths of the 'tracks' as well as the reaction to tiny forces generated by the motors. Acting together, the motors and the asters can form a surprising range of patterns. In our previous work, we have demonstrated how some of these processes are important for the spindle assembly during mouse oocyte maturation [1].

The image

Here, we have extended that work by combining the activity of diffusible double handed motors that can bind two asters at a time, with spring-like anchors in the membrane. The patterns observed 'emerge' from the collective behavior of the parts. At the right range of parameters, we find the star-shaped structures behaving very similar to a flock of birds and start circling the simulated cell when anchors are diffusive. Simulations were run using Cytosim, an agent-based simulator of cytoskeletal mechanics [2].

References

[1]   Khetan N, Athale CA, A motor-gradient and clustering model of the centripetal motility of MTOCs in meiosis I of mouse oocytes, PLoS Comp Biol. 12(10): e1005102, 2016.

[2]   Nédélec F, Foethke D, Collective Langevin dynamics of flexible cytoskeletal fibers, New J Phys. 9: 427, 2007.

© Springer Nature Switzerland AG 2020
F. Matthäus et al. (eds.), *The Art of Theoretical Biology*, https://doi.org/10.1007/978-3-030-33471-0_43

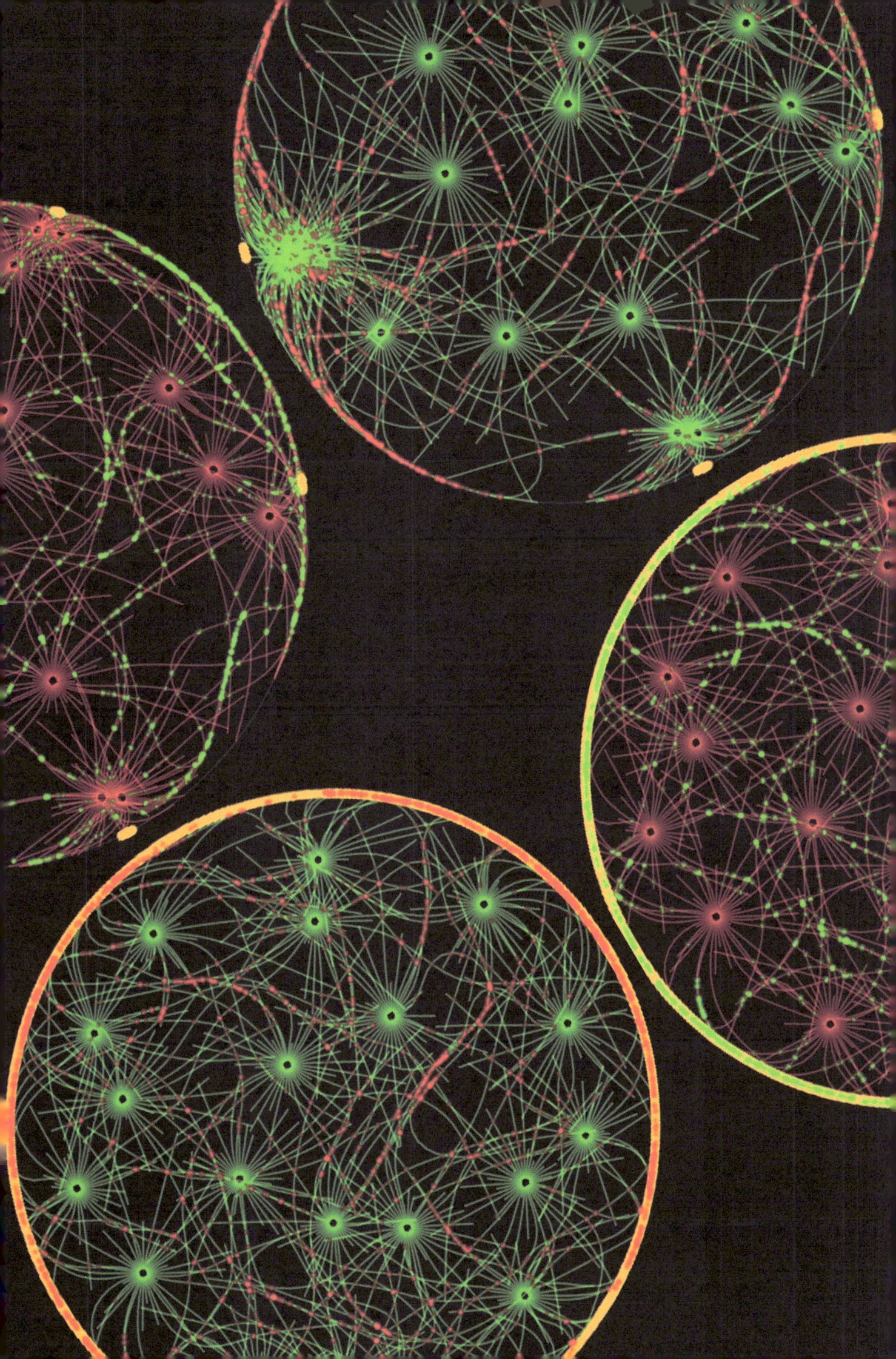

A Mosaic of Cancer and Liver Tissue

By Hermann B. Frieboes, Jessica L. Sparks, Shannon M. Mumenthaler & Paul Macklin

The research story

Cancer cells enter the bloodstream, spread throughout the body, and colonize new tissues. Many cancers spread to the liver, but it's not fully understood why this happens. Is it an easy environment to conquer? Or is it simply a convenient target because all blood circulates through the liver? Once cancer cells arrive in the liver, how do they clear out the liver inhabitants to move in? Our team built a computer model to explore these questions.

The image

In this image, cancer cells (blue and green dots) arrive in the blood, lodge in the liver (tan dots), and seed new tumors. As the new tumors grow, they squeeze the liver cells (strained cells are colored magenta). Eventually, this disruption kills the liver cells, freeing up space for tumor expansion. The cancer cells have a weakness: as they disrupt the liver, they disrupt the blood flow that supplies oxygen and nutrients. Their centers develop cores of dead tissue (brown) that gradually disappear, leaving empty crevices (white). However, the outer living tumor rims find plenty of healthy liver tissue to invade, and a mosaic of liver and cancer tissues emerges. We built the computer model by first simulating how blood flows through the liver and delivers oxygen to the tissue. We then used PhysiCell [1] (a free simulation toolkit we created) to build a computer model of each cell, and to test new ideas on how cancer and liver cells live and die in this constant battle for space and supplies needed for survival.

Reference

[1] Ghaffarizadeh A, Heiland R, Friedman SH, Mumenthaler SM, Macklin P, PhysiCell: an open source physics-based cell simulator for 3-D multicellular systems. PLoS Comput. Biol. 14(2):e1005991, 2018.

© Springer Nature Switzerland AG 2020
F. Matthäus et al. (eds.), *The Art of Theoretical Biology*, https://doi.org/10.1007/978-3-030-33471-0_44

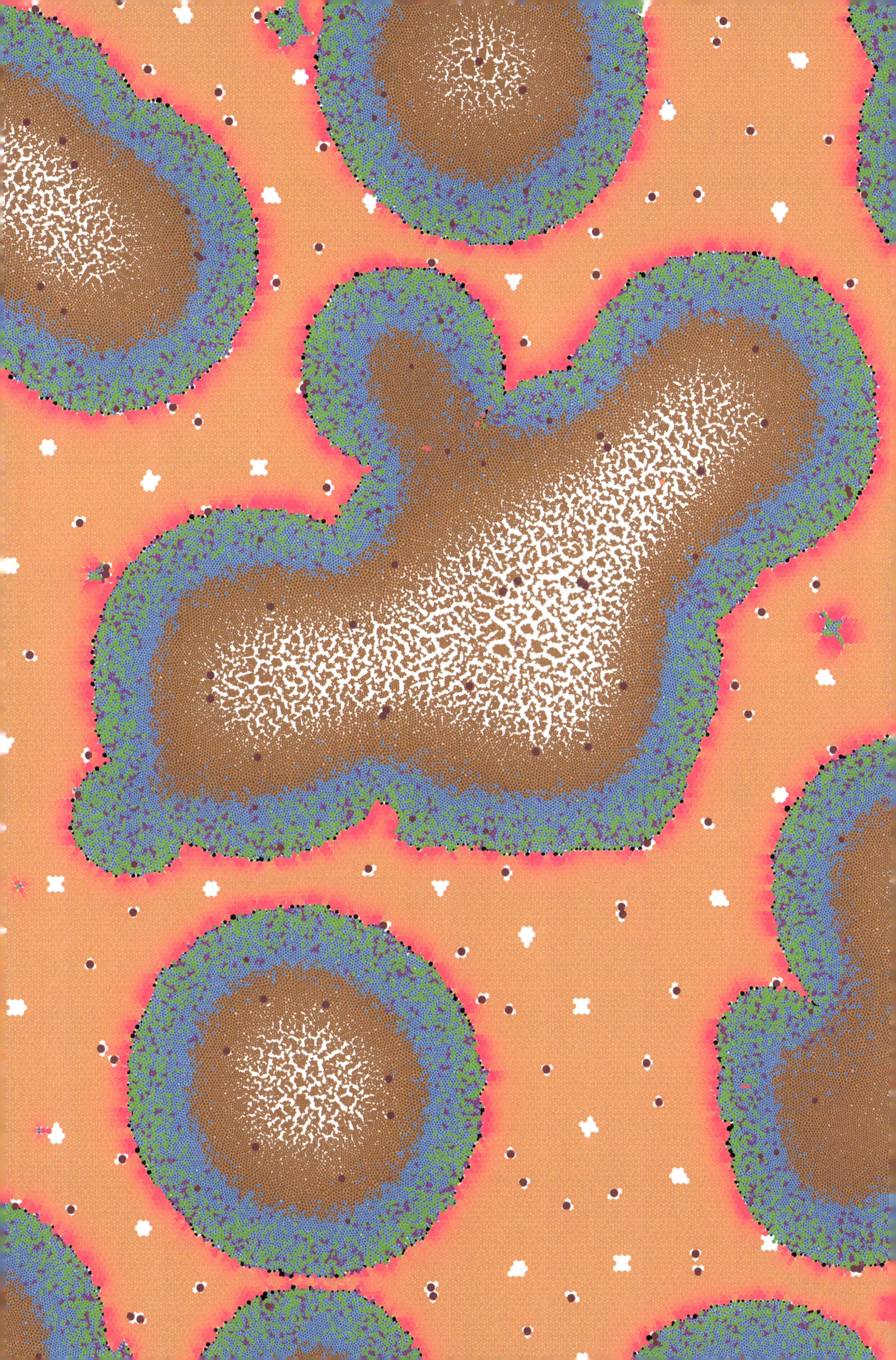

Cell Firework

By Gaëlle Letort

The research story

Microtubules are long and stiff filaments that span the entire cell, providing tracks for intracellular trafficking. Their spatial and temporal distribution plays thus an essential role in regulating cell polarity, shape and other processes. Hundreds of microtubules are generated and organized by the centrosome, the hub of the cell. Its position controls the repartition of the microtubules, and thus the internal organization. In turn, microtubules undergo forces all along their length, by interaction with other proteins, organelles, membranes etc. The integration of all those forces transmitted to the centrosome can move it to respond to the cell's state or shape. Therefore, knowing what determines the position of the centrosome is central in understanding cell processes. To explore the effect of different (off-) centering forces, numerical representations of different identified actors (centrosome, microtubules, cell shape, molecular motors ...) were developed (Cytomorpholab, Grenoble) based on their biological descriptions. Numerical simulations were then used to test how centrosome positioning is affected by different configurations of these factors [1]. This allowed us to identify conditions for robust or sensitive centering, which can be applied to explain or predict different biological situations.

The image

This image is a result of a simulation of microtubules that are radiating out of the centrosome, performed with the Cytosim software. Here, microtubules, shown as red cylinders, are generated from the centrosome (grey sphere) and confined inside a 3D sphere representing the shape of the cell (not visible). Contrast, luminosity and colours levels were modified with Gimp for visual impact.

Reference

[1] Letort G, Nédélec F, Blanchoin L, Théry M, Centrosome centering and decentering by microtubule network rearrangement, MBOC, 2016.

© Springer Nature Switzerland AG 2020
F. Matthäus et al. (eds.), *The Art of Theoretical Biology*, https://doi.org/10.1007/978-3-030-33471-0_45

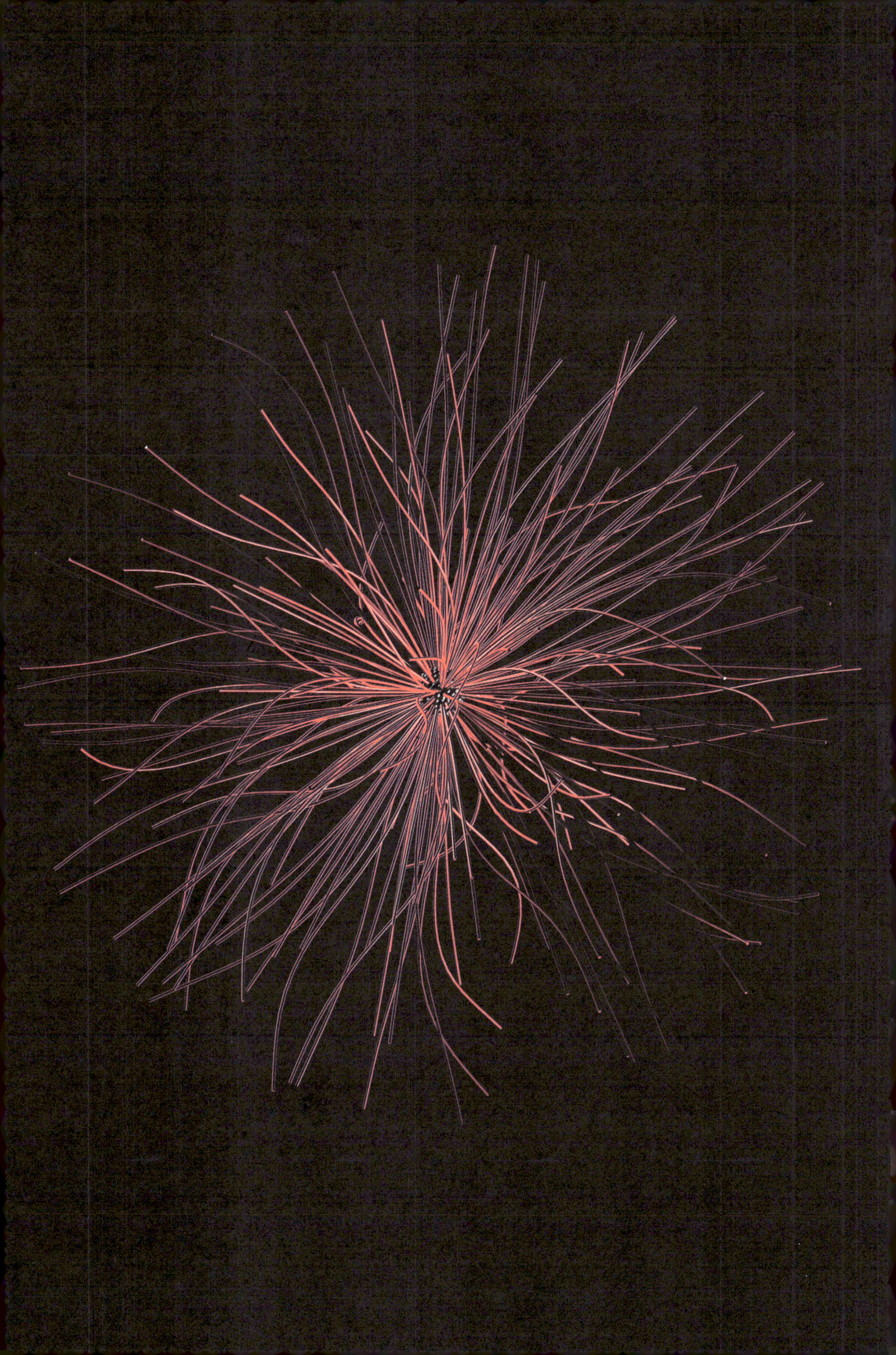

Pulled in Line

By Elisabeth Rens & Roeland Merks

The research story

The formation of blood vessels plays a pivotal role in tumour growth. "Pulled in line" shows a simulation of how a tumour could attract blood vessels by pulling on the extracellular matrix. Many cell types in our body are able to apply forces on their surroundings. The gel-like structure surrounding cells and tissues is called the extracellular matrix. It becomes stressed due to the pulling forces of cells. Because cells can feel and respond to these stresses, cells can communicate their position to other cells.

The image

In this simulation we started out with the circular cluster of cells on the top left. We then randomly distributed the other cells around it. In this graphic the cells are transparent and they have black boundaries. The colors indicate the stress in the extracellular matrix. Colors are linearly interpolated in CMYK color space between Ivory Buff for zero stress, Light Greyish Olive for intermediate stress and Spinel Red for high stress. This is color combination 184 in "A Dictionary of Color Combinations" (Seigensha Publishing). We then let cells extend into areas of high stress. This resulted in the pattern that is shown in this graphic: cells line up with each other and the lines connect to the cluster. In our research [1,2] we have used this model to analyse how the extracellular matrix can mediate the cellular interactions that help them organise into blood-vessel-like cellular networks. We have found that networks can form if the extracellular matrix is not too stiff and not too soft, in agreement with experimental observations.

References

[1] van Oers RFM, Rens EG, LaValley DJ, Reinhart-King CA, Merks RMH, Mechanical cell-matrix feedback explains pairwise and collective endothelial cell behavior in vitro, PLoS Computational Biology 10: e1003774, 2014.

[2] Rens EG, Merks RMH, Cell contractility facilitates alignment of cells and tissues to static uniaxial stretch, Biophysical Journal 112: 755–766, 2017.

© Springer Nature Switzerland AG 2020
F. Matthäus et al. (eds.), *The Art of Theoretical Biology*, https://doi.org/10.1007/978-3-030-33471-0_46

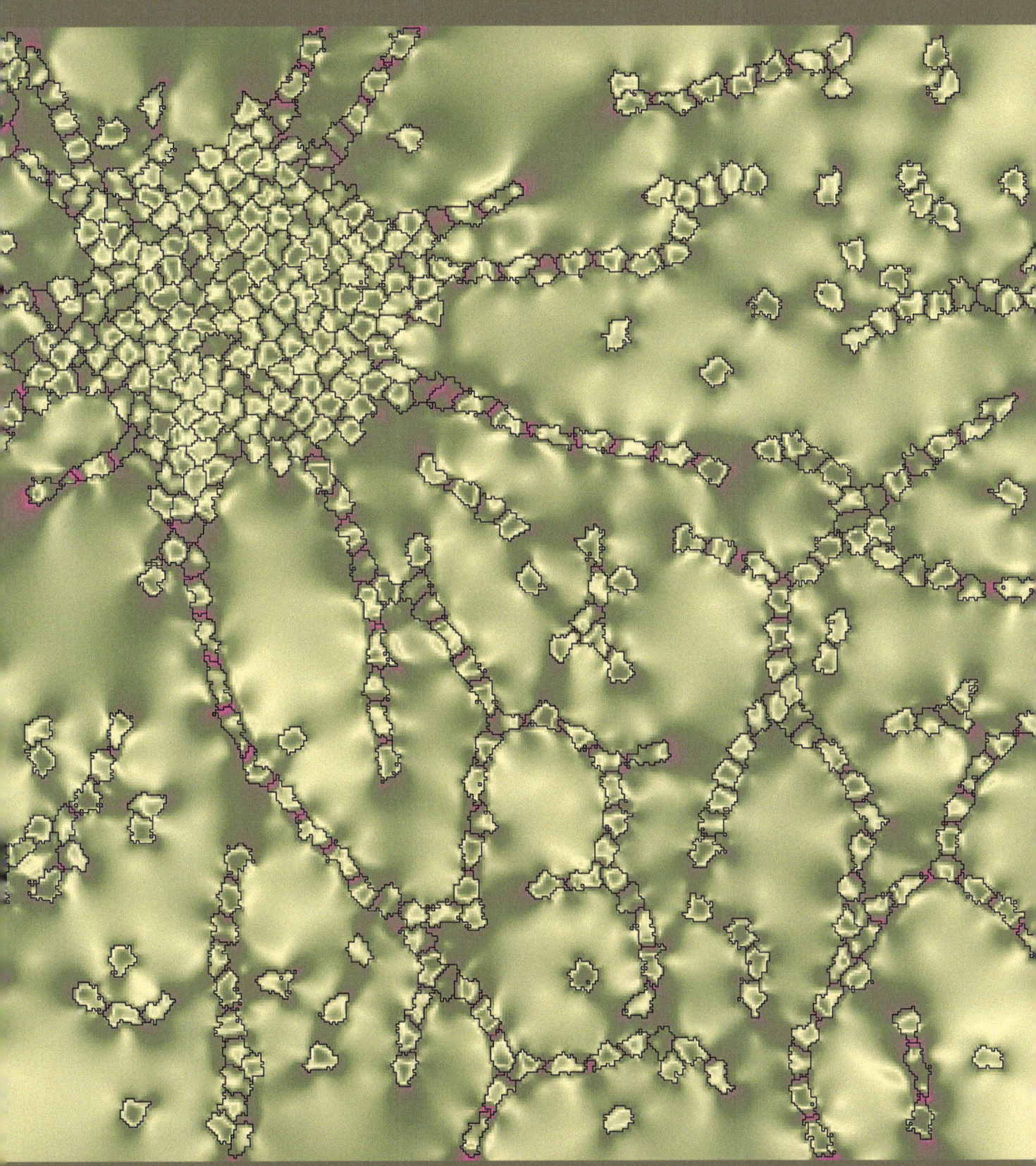

Convergence

The research story

Cell migration, division, morphogenesis and many other cellular processes rely on actin filaments. Being short and flexible polymers, they contribute to cell processes through collective action. They organize into networks of specified structures to be able to perform different functions (e.g. pushing the membrane, forming a contractile ring to separate the two daugther cells). Architecture and composition of these networks determine their efficiency to fullfill variable roles. In particular, molecular motors can associate to these networks to trigger a contractile behavior. To understand network contractile responses, in-vitro reconstitutions were combined with numerical simulations (Cytomorpholab, Grenoble).

The image

This image illustrates numerical simulations performed on three networks with different architectures but based on the same components, used in Ennomani et al, 2016 [1]. One "circle" corresponds to the behavior of one network. One white point in the image is one pointed end (extremity of actin filament) position at a given time point. Initially, networks are circles of the same radius (outer radius). Upon addition of motors, two networks contract: their pointed ends are brought together. Due to its organization, the top network does not deform but keeps its initial radius. Simulations were performed with Cytosim software, from which positions of the pointed end of the filaments at each time point were imaged. The projections of all time points together and combination of the three results in the single resulting image were done with ImageJ.

Reference

[1] Ennomani H, Letort G, Guérin C, Martiel JL, Cao W, Nédélec F, De La Cruz EM, Théry M, Blanchoin L, Architecture and connectivity govern actin network contractility, Current Biology 26(5): 616–26, 2016.

© Springer Nature Switzerland AG 2020
F. Matthäus et al. (eds.), *The Art of Theoretical Biology*, https://doi.org/10.1007/978-3-030-33471-0_47

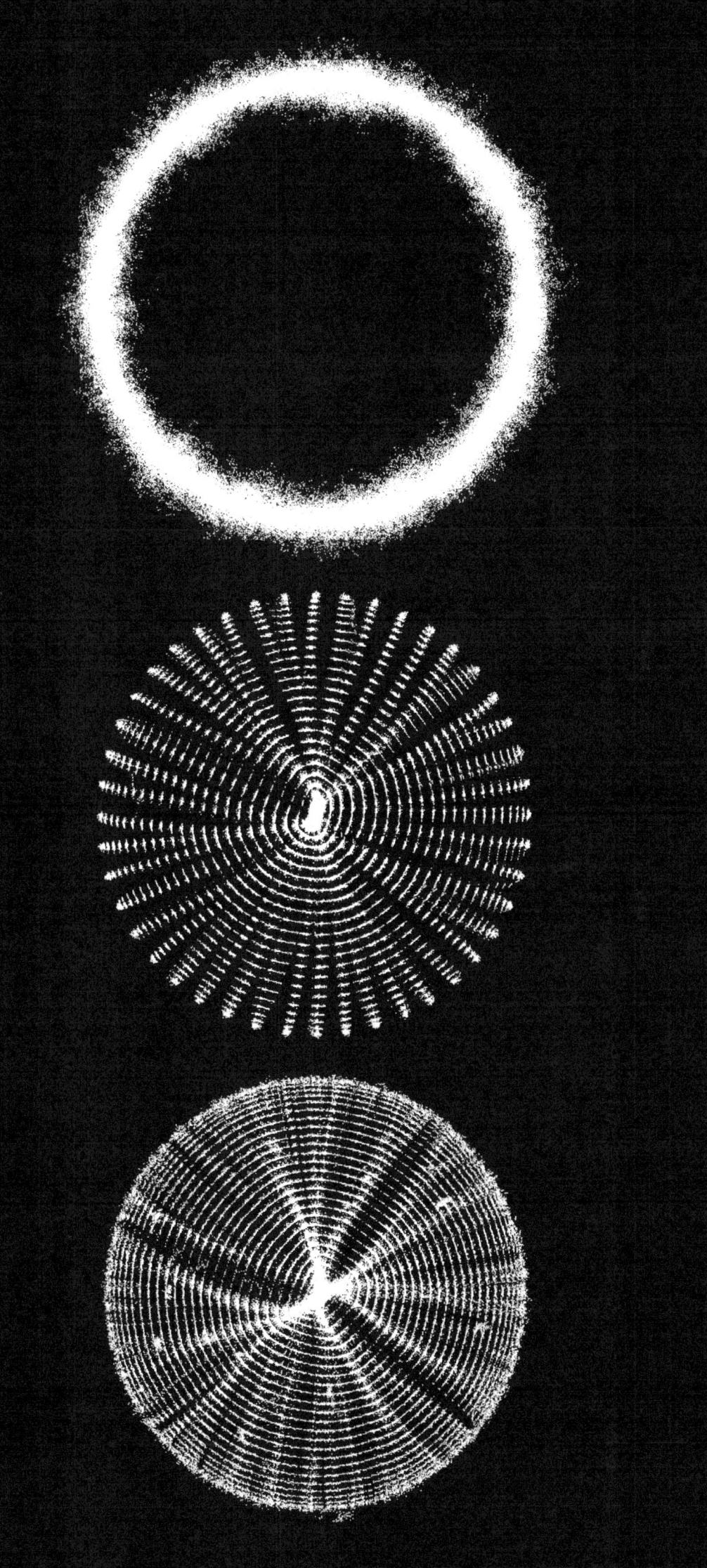

Arctic Breeze

By Laura Fischer & Franziska Matthäus

The research story

Many bacteria are capable of sensing changes in the concentration of chemical substances. They use this capability to move towards nutrient sources or away from toxic substances – a process called chemotaxis. Cells of higher organisms can also detect in which direction a concentration increases, simply by comparing the concentrations at different positions of the cell membrane. But bacteria are much smaller, and concentrations at opposite cell ends do not differ much. Bacteria of the type Escherichia coli therefore use a type of "sensory memory". Hereby, a chemical signaling system inside the cell stores information about past concentrations. While swimming they compare the past value to the present concentration and decide if the direction is favorable. The chemotaxis system of E. coli bacteria is one of the simplest and best understood model systems – both on the experimental as well as on the theoretical side.

The image

In a study we investigated how a population of E. coli bacteria moves in a one-dimensional domain in which nutrients are constantly produced and consumed by the bacteria. Starting from a homogeneous distribution the bacteria accumulated in some region of the domain, reduced the concentration of the nutrients and then moved collectively to another domain. While the bacteria were feeding at one position, the nutrients in the deserted area were restored, and then attracted the population again. The image shows the distribution of the bacteria in the one-dimensional domain and its temporal dynamics.

F. Matthäus et al. (eds.), *The Art of Theoretical Biology*, https://doi.org/10.1007/978-3-030-33471-0_48

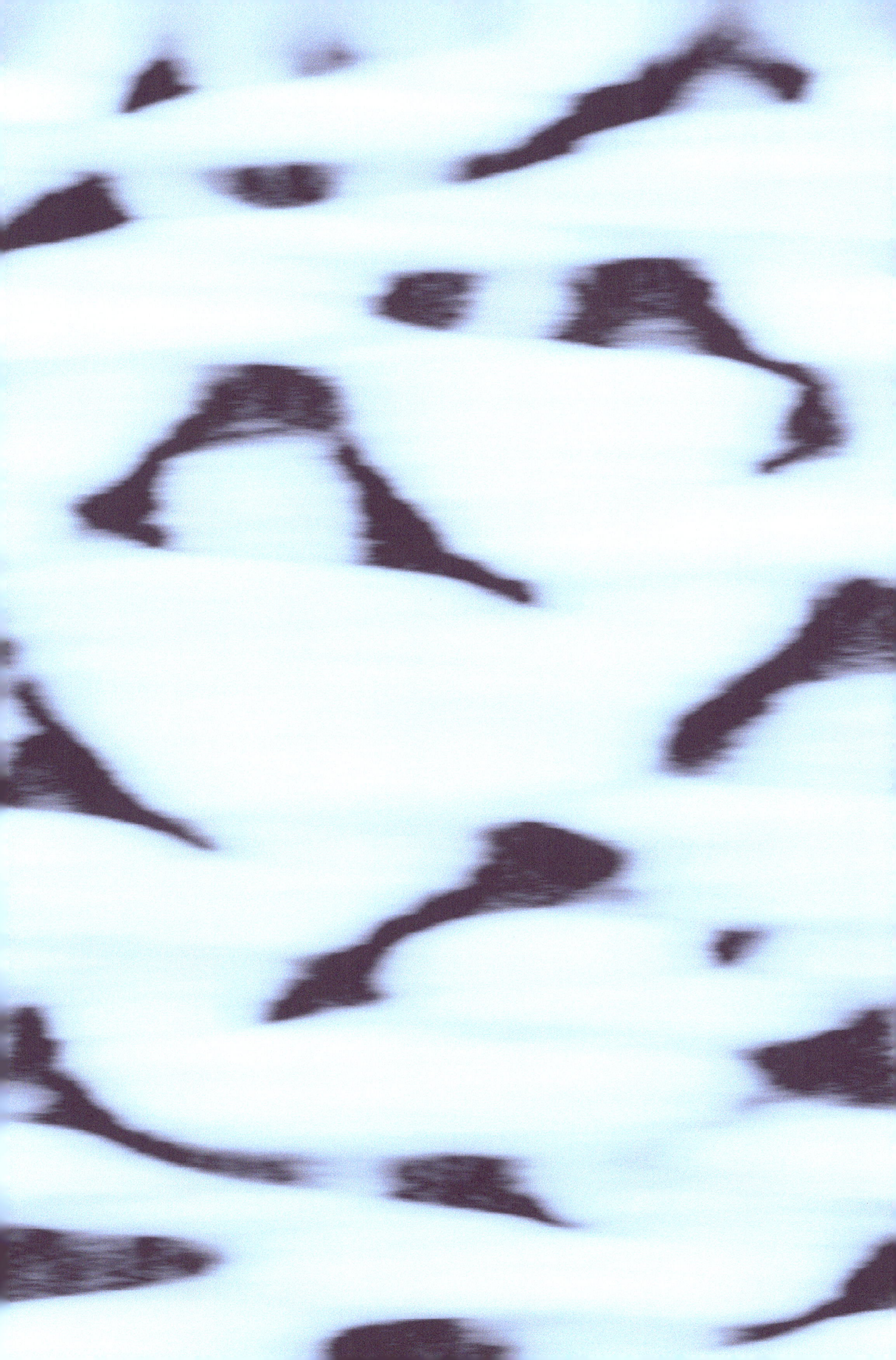

Extracellular Galaxies

By Russell Rockne & Michael Barish

The research story

Recent advances in immunotherapy for cancer have piqued our interest in the dynamic interplay between the immune system and the cancer microenvironment. In order for immunotherapy to be effective, the cat and mouse game between the immune cells and the cancer cells must be in the immune cell's favor. But the cancer cells are very good at hiding – in plain sight – from the immune cells. The research question is to understand how to eradicate the cancer cells by training the immune system to better seek and destroy the cancer cells by removing the cancer cell's camouflage.

The image

Here we see a pseudo-colored image of the extracellular, extravascular space within a small region of a theoretical brain tumor which produces a favorable environment for the immune system to operate. The center of the galaxy is full of dead cancer cells. This Extracellular Galaxy is full of potential for the immune system to effectively eradicate the cancer cells, as well as provide a window into the mysterious and dynamic space-time of the cancer microenvironment, in which the cancer cells invoke a Cloak of Invisibility to evade the immune system. We beg to ask the question: is the cancer warping its local biological space-time continuum? Is the Extracellular Galaxy expanding or contracting? One may never know.

© Springer Nature Switzerland AG 2020
F. Matthäus et al. (eds.), *The Art of Theoretical Biology*, https://doi.org/10.1007/978-3-030-33471-0_49

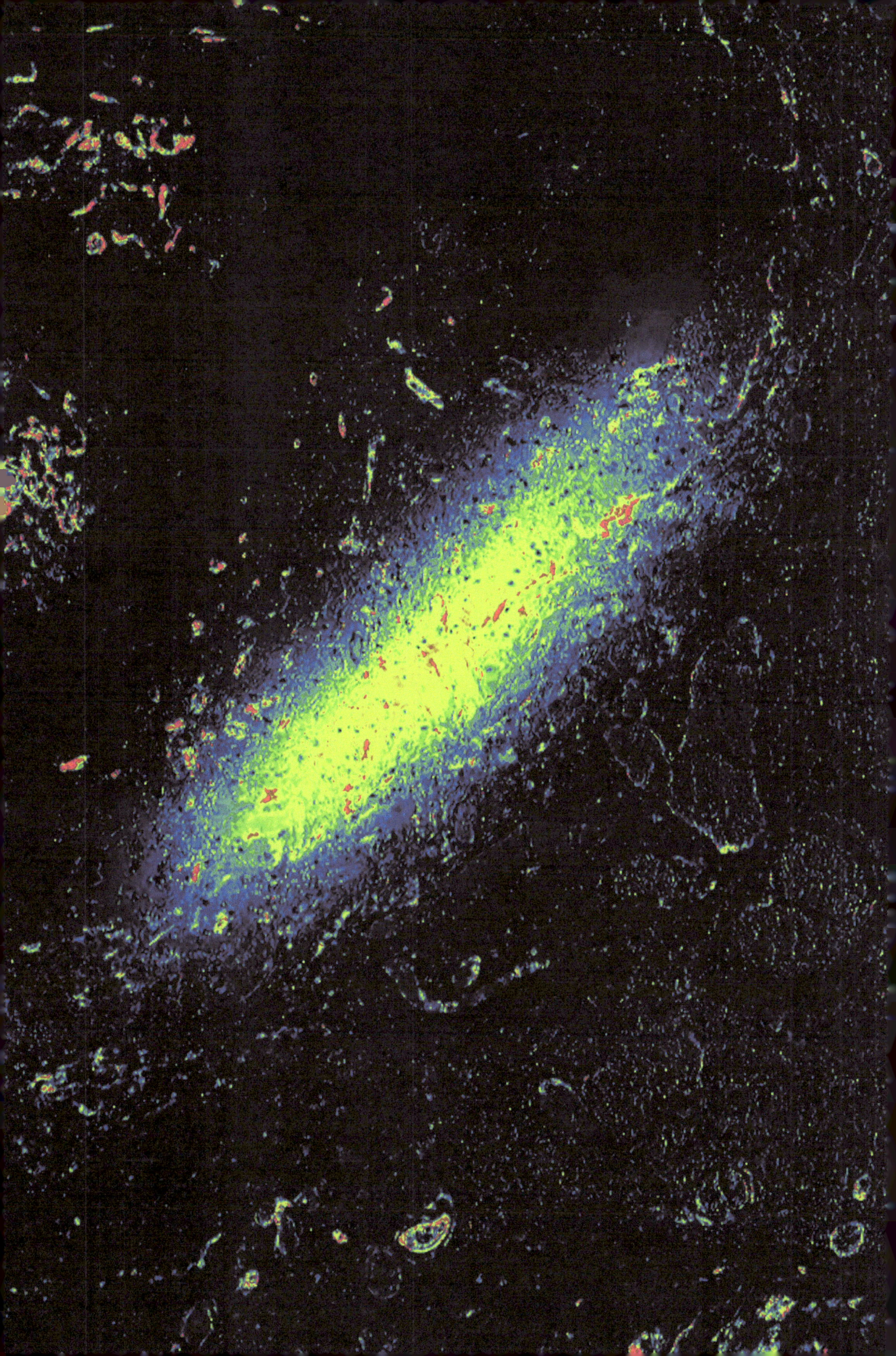

Oriental Landscape Painting by Predator Species

By Yong-Jung Kim

The research story

Steele [1] conjectured that Lotka-Volterra predator-prey equations with diffusion would produce patterns of plankton patchiness. However, mathematicians showed that the solution eventually converges to a constant and there are no persistent patterns of the system. This observation motivates researchers to introduce various nonlinear functional responses and modify the equations. These modified equations successfully provide persistent patterns. The distribution landscape shows areas where the population flourishes and areas where it is not present. The figure is obtained by adding the death term (or the Allee effect) which gives finite time local extinction.

The image

Mountain and water are two main topics of oriental landscape paintings. The figure illustrates the density distribution of a predator species, which resembles oriental landscape paintings. Turing patterns are static images usually obtained when the diffusivity of two species has a big difference. The figure is such a case when the diffusivity of the predator species is much smaller than that of the prey species. As a result, the predator landscape is very spiky and exotic. The prey landscape is smoother and relaxing, which can be found at http://amath.kaist.ac.kr/predatorprey/. However, this image is a dynamic pattern and keeps evolving. See [2] to find a Turing pattern of predator-prey equations.

References

[1]   Steele JH, Spatial heterogeneity and population stability, Nature 248:83, 1974.

[2]   Choi J, Kim Y-J, Predator-prey equations with constant harvesting and planting, Journal of Theoretical Biology 458: 47–57, 2018.

© Springer Nature Switzerland AG 2020
F. Matthäus et al. (eds.), *The Art of Theoretical Biology*, https://doi.org/10.1007/978-3-030-33471-0_50

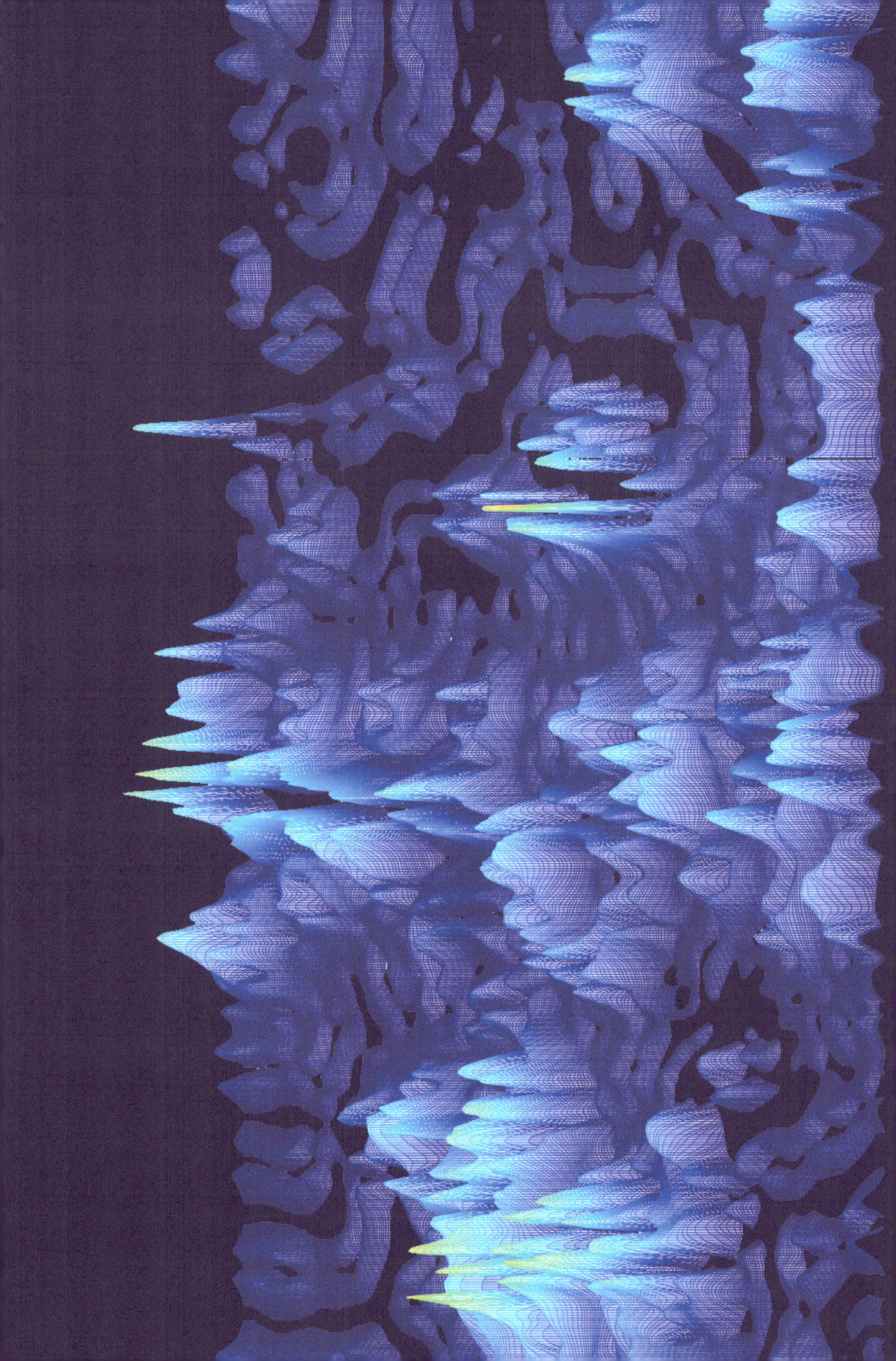

Life is Lived on the Edge

By Glenn Webb

The research story

Mathematical models of complex dynamic biological processes sometimes involve systems of partial differential equations for production, growth, decay, interaction, and spatial movement. An example is a nonlinear partial differential equations model for a very small tumor invading surrounding tissue. The outward growth of the tumor results from haptotaxis, which is the directional movement of cells up-gradient a chemoattractant to adhesion sites in the surrounding tissue. The model equations involve five key components of the tumor growth process: proliferating tumor cells (cells progressing through the cell cycle to division), quiescent tumor cells (cells not progressing through the cell cycle, but capable of returning to cell cycle progression), surrounding tissue macromolecules (the cellular environment of the tumor), degradative enzymes (which limit tumor growth), and oxygen (which supports tumor growth).

The image

The images in the figure, obtained from numerical simulation of the model, track the growth of the tumor at six time points. The tumor initially consists of proliferating cells. At time point 0, the tumor is spatially homogeneous and very small. At later times the tumor grows outwardly in an irregular way. As time advances, the interior section of the tumor becomes composed mostly of quiescent cells (black), with more and more proliferating cells (red) at the outer edges.

References

[1] Dyson J, Villella-Bressan R, Webb GF, An age and spatially structured model of tumor invasion with haptotaxis II, Math. Pop. Studies 15: 1–23, 2008.

[2] Walker C and Webb GF, Global existence of classical solutions for a haptotaxis model, SIAM J. Math. Anal. 38: 1694–1713, 2007.

© Springer Nature Switzerland AG 2020
F. Matthäus et al. (eds.), *The Art of Theoretical Biology*, https://doi.org/10.1007/978-3-030-33471-0_51

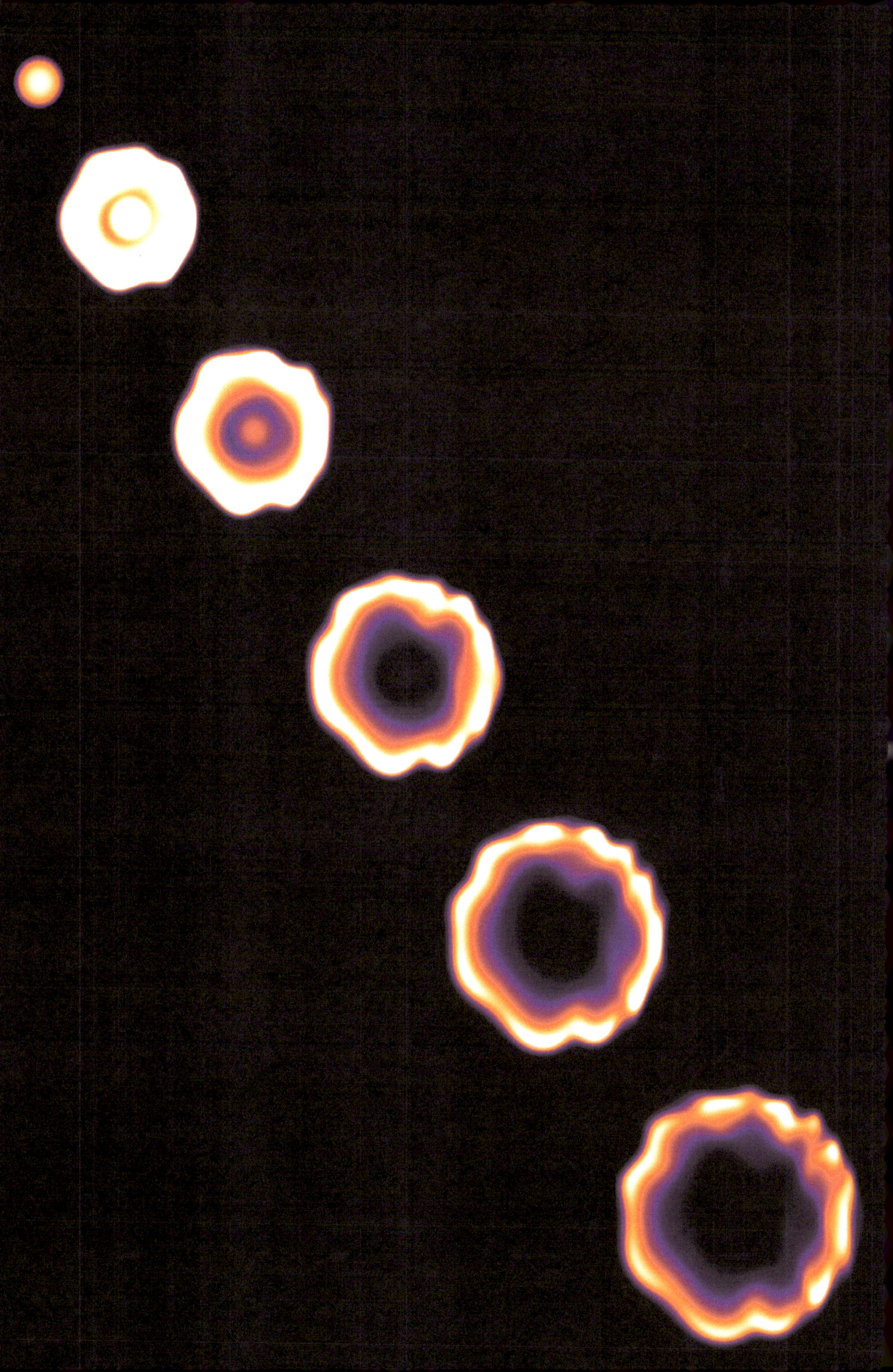

Cellular Swarms in Cellular Automata

By Andreas Deutsch

The research story

Deciphering organizational principles of multi-cellular systems is crucial for understanding key processes in biological development, regeneration, and disease dynamics. Multicellular systems are driven by migration, differentiation and proliferation of individual cells. Meanwhile various cell-based mathematical models have been introduced to analyze such systems. These models are typically either very detailed biophysically, hampering computational and mathematical analysis, or, they are very abstract, guided by phenomenological laws. Biological lattice-gas cellular automata (BIO-LGCA), a particular type of cellular automata, provide a mesoscopic computationally efficient modeling framework for multicellular systems, allowing to analyze the coupling of micro/intracellular and macro/tissue scale phenomena [1]. Their rules can be derived from microscopic biophysical laws [2]. The BIO-LGCA model is based on a statistical description of the microenvironment and permits multiscale analysis.

The image

The figure shows a snapshot of a simulation based on a BIO-LGCA that we have introduced as a model for random walkers with interactions favoring local alignment and leading to collective motion and swarming. The degree of alignment is controlled by a sensitivity parameter. A dynamical phase transition exhibiting spontaneous breaking of rotational symmetry occurs at critical parameter values. To analyze the model, mean-field approximations have been derived that predict the phase transition as a function of sensitivity and density. Different colors encode different cell orientations. The figure indicates formation of aligned clusters.

References

[1] Deutsch A, Dormann S, Cellular Automaton Modeling of Biological Pattern Formation: Characterization, Applications, and Analysis, Birkhäuser, Boston, 2018.

[2] Nava-Sedeno JM, Hatzikirou H, Peruani F, Deutsch A: Extracting cellular automaton rules from physical Langevin equation models for single and collective cell migration, J. Math. Biol. 75(5): 1075–1100, 2017.

© Springer Nature Switzerland AG 2020
F. Matthäus et al. (eds.), *The Art of Theoretical Biology*, https://doi.org/10.1007/978-3-030-33471-0_52

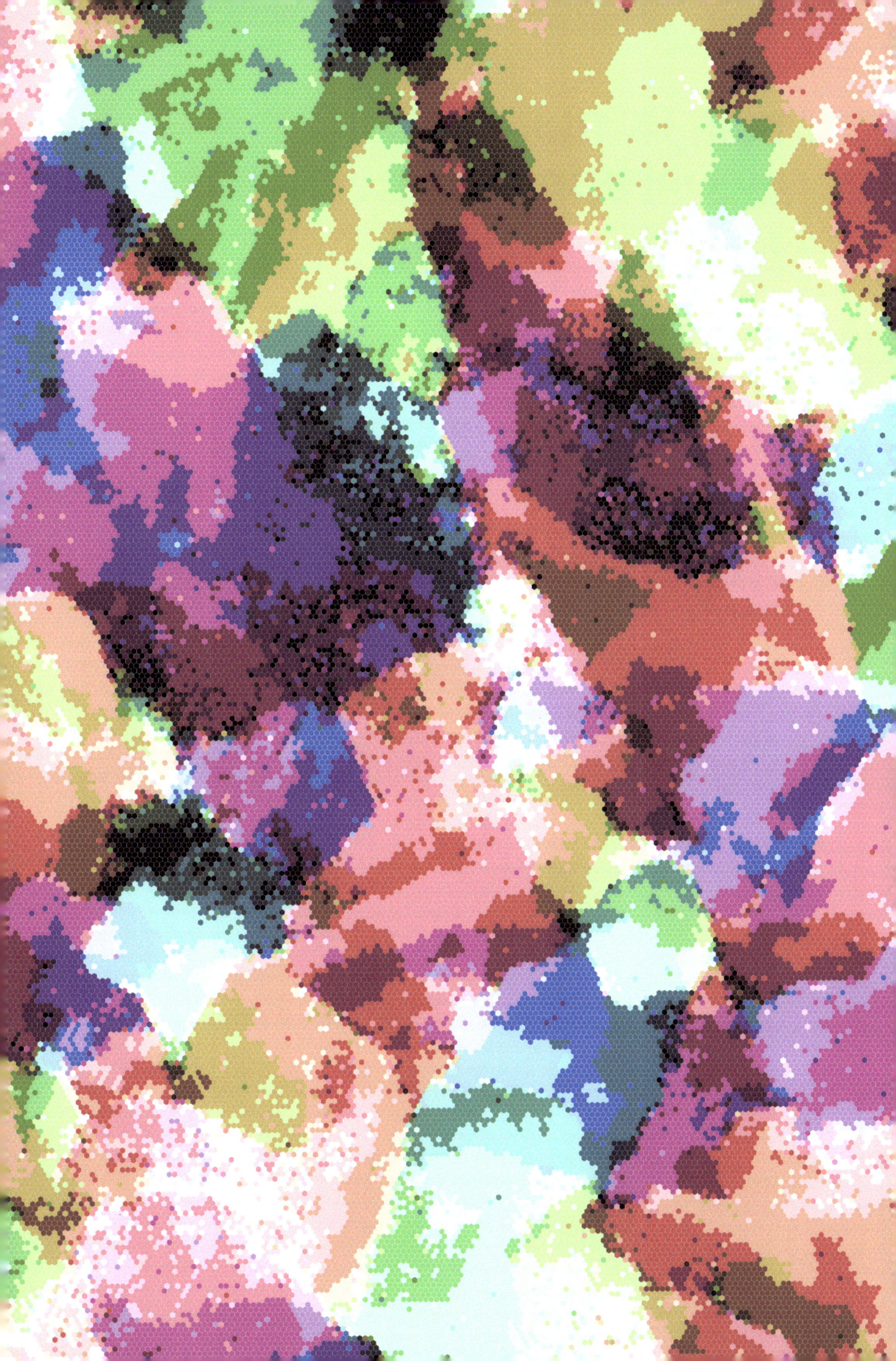

Bumps, Ridges, and no Flows in Vein

By Nicholas Battista & Laura Miller

The research story

In a developing vertebrate heart, the heart does not simply beat just to transport blood, nutrients, and oxygen, but it beats to aid in the formation of the heart itself! Our hearts begin as a straight, valveless tube and morph into a multi-chambered valvular pumping system during development. A well-choreographed dance takes place between the shape of the heart and the blood being pumped through it. As the blood is driven against the chamber walls, little cellular hair-like sensors along the heart walls feel the force of the blood, take that signal, and delegate a list of orders to spark other cellular processes to transform the heart through various geometrical changes. During this whole operation, the composition of heart walls themselves undergo complex morphological changes themselves, where small bumps and ridges, called trabeculae, push out and form. It's been suggested that these trabeculae coordinate how much force the heart walls feel from the blood [1].

The image

This image depicts a computational model of a developing heart, highlighting the interplay between these trabeculae, found along the edge of the ventricle, and blood cells as they are pushed through. The dark and light regions illustrate areas where the blood forms vortices as the blood is driven up into the ventricle from the atrium. If any of these processes are slightly off, the heart will not develop correctly and a congenital heart defect arises.

Reference

[1] Battista NA, Lane AN, Liu J, Miller LA, Fluid dynamics in heart development: effects of trabeculae and hematocrit, Math. Med. and Biol. 35(4): 493–516, 2018.

F. Matthäus et al. (eds.), *The Art of Theoretical Biology*, https://doi.org/10.1007/978-3-030-33471-0_53

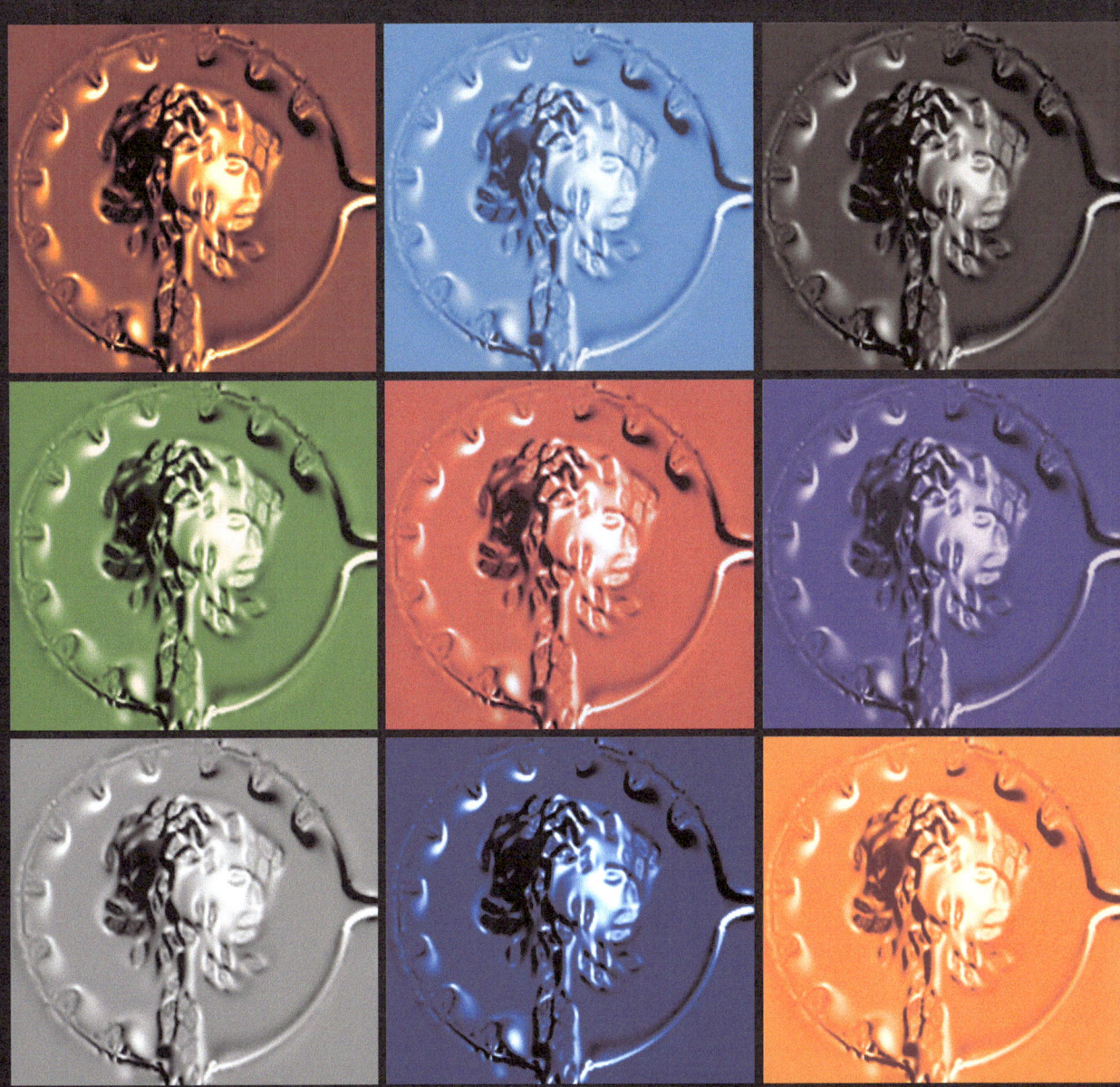

Growing Orbs / Mingled Metabolism

By Huaming Yan, Anna Konstorum & John S Lowengrub

The research story

Many solid tumors exhibit a striking range of diverse subtypes of cancer cells that help the tumor survive harsh circumstances such as the limited availability of nutrients, an attacking immune system or chemotherapy treatment. However, the actual mechanisms that generate this diversity are poorly understood. Here, we use multiscale mathematical modeling to investigate the emergence and consequences of non-genetic heterogeneity. We focus on colon cancer, which is a heterogeneous and complex disease.

The image

We developed a three-dimensional mathematical model to simulate the growth of colon cancer organoids containing stem, progenitor and terminally differentiated cells, as a model of early (prevascular) tumor growth. Stem cells (SCs, shown in red) produce short-range-acting self-renewal signals (e.g. Wnt) as well as long-range-acting inhibitors (e.g. Dkk), and they proliferate slowly. Committed progenitor cells (CP, shown in green) proliferate more rapidly to produce terminally differentiated cells that do not divide but release signals that act as negative feedback loops on SC and CP self-renewal. The mathematical modeling shows that SCs play a central role in cancer colon organoids. In particular, spatial patterning of the SC self-renewal signal gives rise to SC clusters, which mimic stem cell niches, around the organoid surface and drive the development of invasive fingers at later times. Since many cancers are hierarchically-organized and are subject to feedback regulation, our results suggest that controlling cancer stem cell self-renewal should influence the size and shape of the tumor, thereby opening the door to novel therapies that target feedback inhibition by multiple cancer cell subtypes.

Reference

[1] Yan H, Konstorum A, Lowengrub JS, Three-dimensional spatiotemporal modeling of colon cancer organoids reveals that multimodal control of stem cell self-renewal is a critical determinant of size and shape in early stages of tumor growth, Bull Math Biol 80: 1404, 2018.

© Springer Nature Switzerland AG 2020
F. Matthäus et al. (eds.), *The Art of Theoretical Biology*, https://doi.org/10.1007/978-3-030-33471-0_54

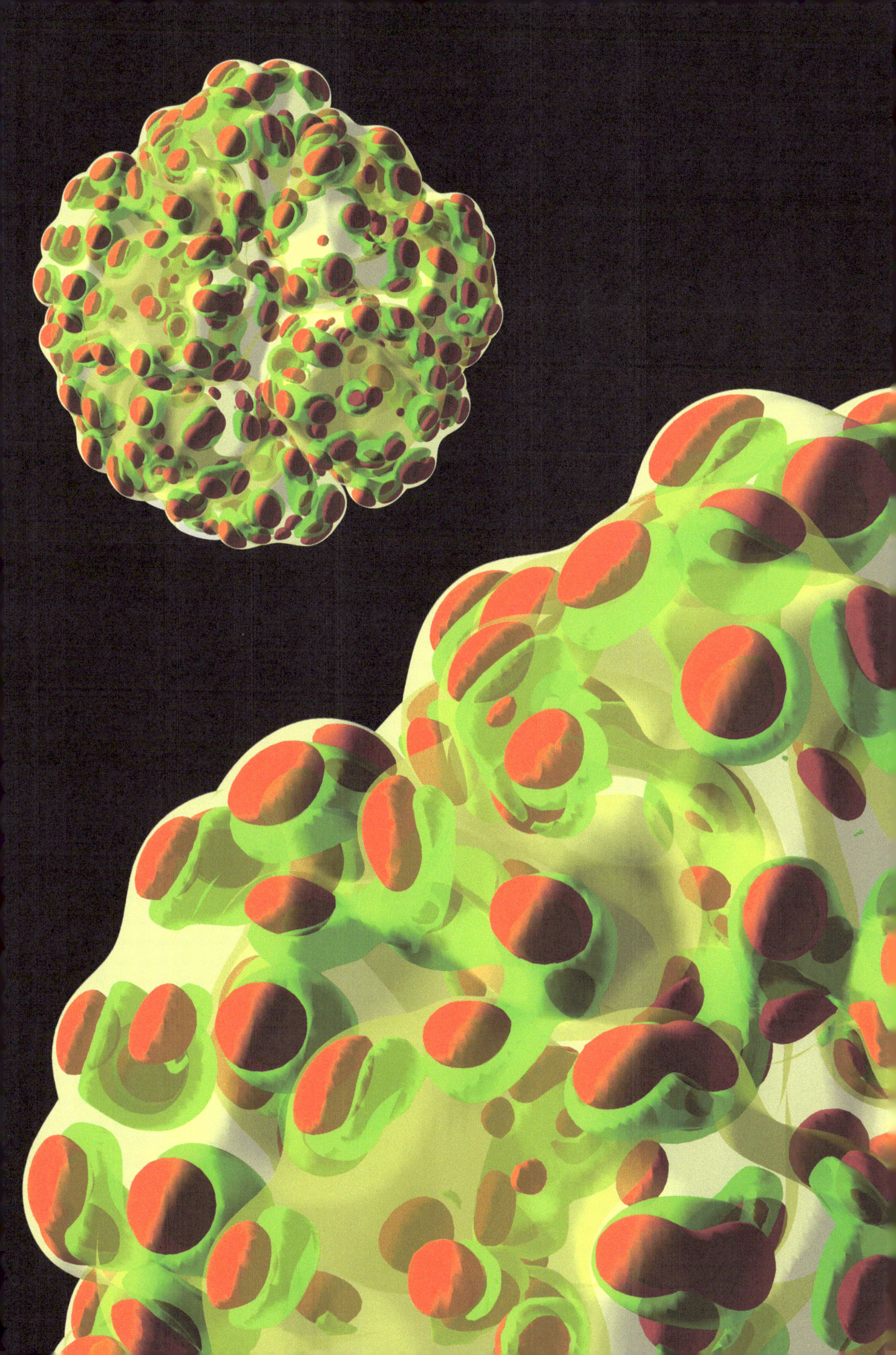

Out of the Comfort Zone

By Jana Lipková, Diego Rossinelli, Petros Koumoutsakos & Bjoern Menze

The research story

Particle methods are a natural way of modelling flow problems. Particles can be seen as objects carrying a physical property of a system, following dynamics of a flow field. To assess an effectiveness and accuracy of a numerical method a following benchmark problem is considered. A sphere of a radius 0.15 and centre at (0.35, 0.35, 0.35) is placed inside a unit cube. The sphere is then deformed by a divergent-free flow field proposed in [1].

The image

The visualisation shows three stages of the sphere's deformation, with the initial state shown in the lower left corner, intermediate state in the middle and final deformed state in the upper right corner. After reaching the final state, the flow deforms the interface back to the initial sphere, following the shown deformed states in a reverse order. Since the deformation field is reversible, it is possible to evaluate the accuracy of a numerical method under complex conditions without knowing the analytical solution. For instance, incapability of the solver to reproduce the initial sphere indicates loss of the mass and dissipation of the method. The 3D visualisation is used to identify sources and locations of such errors. The present visualisation is performed in Volume Perception [2], a volume rendering software using a ray-casting technique. The field is visualised as translucent isosurfaces obtained by pre-integrated volume rendering.

References

[1]   Leveque RJ, High-resolution conservative algorithms for advection in incompressible flow. SIAM Journal on Numerical Analysis 33(2):627–665, 1996.

[2]   Rossinelli D, Multiresolution flow simulations on multi/many-core architectures, PhD thesis, ETH Zurich, 2011.

© Springer Nature Switzerland AG 2020
F. Matthäus et al. (eds.), *The Art of Theoretical Biology*, https://doi.org/10.1007/978-3-030-33471-0_55

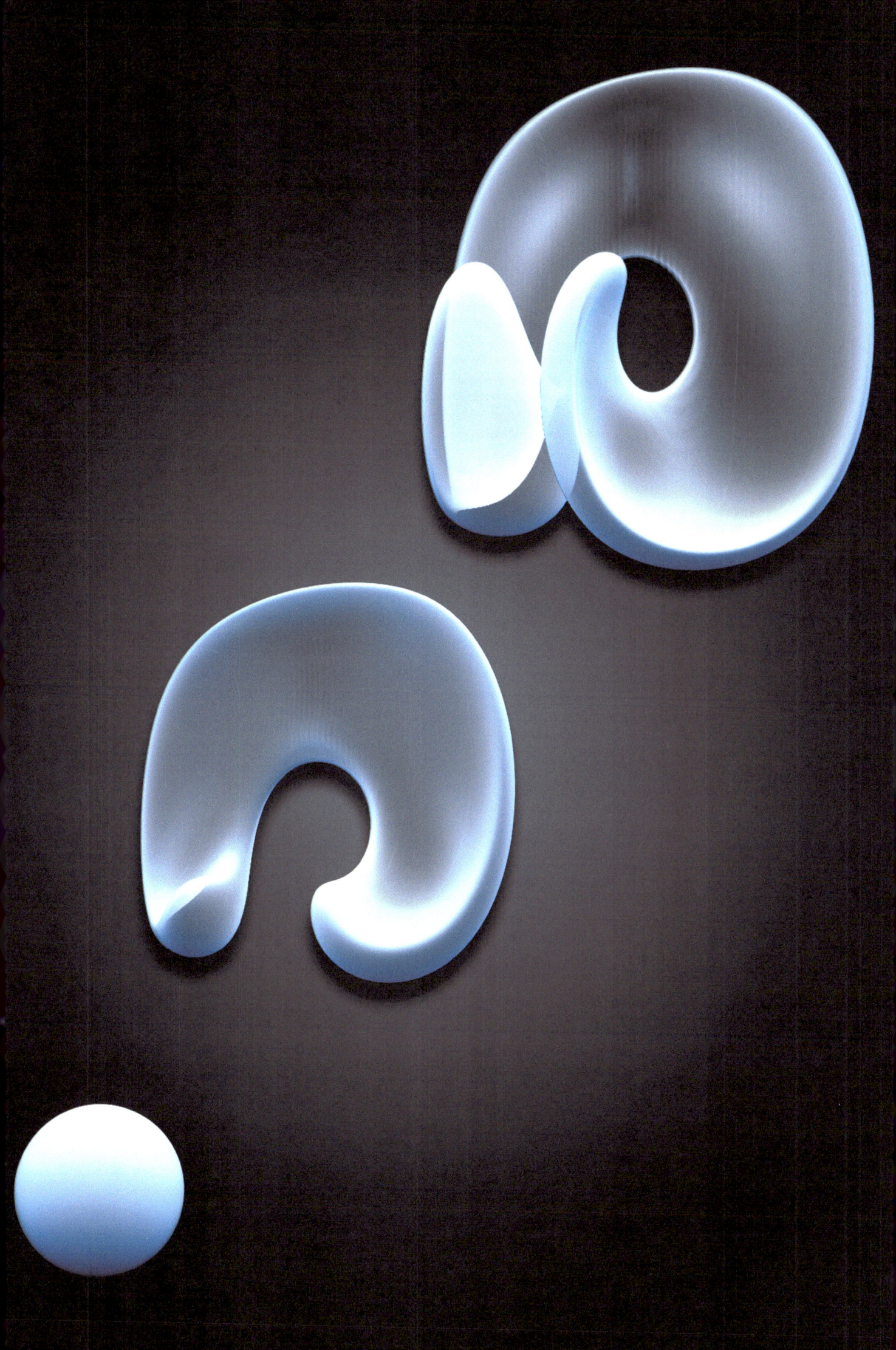

Green Protein Interaction Wheel

By Jens Rieser, Jörg Ackermann & Ina Koch

The research story

Salmonella Typhimurium causes several thousand deaths every year. The first reaction of a cell against such an infection is the posttranslational modification of proteins. A better understanding of this immune response can lead to new drug development or treatment methods. Two of the possible modifications are ubiquitination and phosphorylation of proteins. To put both types of immune reactions together, we selected two datasets from the literature [1,2] and extracted the proteins. For these proteins, we performed a search for all interactions in public databases such as STRING, IntAct or BioGRID. The resulting protein-protein interaction network consists of 18,978 interactions between 1,704 proteins.

The image

This picture illustrates the proteins clustered in different groups dependent on their type of modification. Five clusters contain only phosphorylated proteins, five only ubiquitinated and one cluster includes all proteins with both posttranslational modifications. The order of the clusters makes it possible to investigate the influence of a highly modified group to other groups of proteins. Although the different groups can be separately analyzed, clusters with both modifications are of great interest. Beside this representation of the network, there are different other possibilities for further analysis like the basic network properties or a GO-enrichment for functional analysis.

References

[1] Fiskin E, Bionda T, Dikic I, Behrends C. Global analysis of host and bacterial ubiquitinome in response to Salmonella typhimurium infection, Molecular Cell 62:967–981, 2016.

[2] Rogers LD, Brown NF, Fang Y, Pelech S, Foster LJ, Phosphoproteomic analysis of Salmonella-infected cells identifies key kinase regulators and SopB-dependent host phosphorylation events, Science Signal 4:191, 2011.

© Springer Nature Switzerland AG 2020
F. Matthäus et al. (eds.), *The Art of Theoretical Biology*, https://doi.org/10.1007/978-3-030-33471-0_56

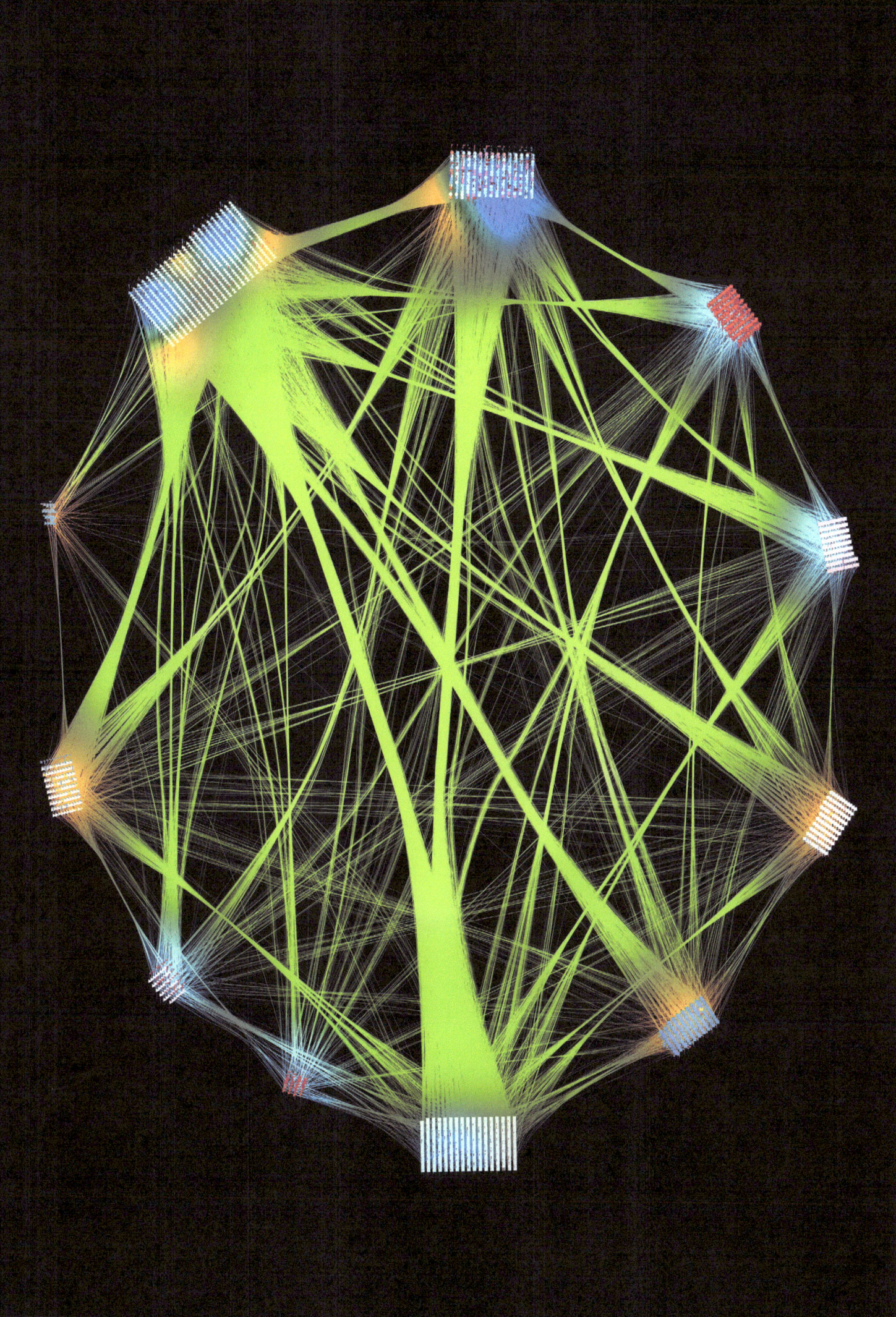

Vincent van Gogh's Autocatalysis

By Amelia Palermo, Lorenzo Casalino & Giulia Palermo

The research story

RNA is a fundamental molecule that codes for protein and controls gene expression, playing a fundamental role in many biological processes. The genetic information contained in premature messenger RNA (mRNA) is cleared of its non-coding sections, known as introns, to be converted to proteins. In several simpler organisms, this key process is carried out by group II introns, enzymes entirely made up of RNA that are able to self-cleave and remove themselves from the mRNA filament, thereby enabling RNA maturation and protein expression. Group II introns employ magnesium atoms to perform the "self cleavage", which help the enzyme in acting like a "double scissors" to cleave itself.

The image

In this painting, two scissors are used to represent the splicing mechanism operated by Group II intron ribozyme. This artistic rendition is an original handmade painting of Amelia Palermo (TSRI, La Jolla), digitally designed and manipulated by Lorenzo Casalino (UC San Diego), over an original idea of Giulia Palermo (UC Riverside). This painting has been inspired by recently published research on the mechanism of RNA splicing [1,2]. This artistic picture has received the first place prize at the 61st Biophysical Society Art of Science Image Contest, 2017, February, New Orleans. It has been selected as a Cover for the May 2019 issue of the Journal of Structural Biology.

References

[1] Casalino L, Palermo G, Spinello A, Rothlisberger U, Magistrato A, All-atom simulations disentangle the functional dynamics underlying gene maturation in the intron lariat spliceosome. Proc Natl Acad Sci USA 115: 6584–6589, 2018.

[2] Casalino L, Palermo G, Rothlisberger U, Magistrato A, Who activates the nucleophile in ribozyme catalysis? An answer from the splicing mechanism of group II introns. J. Am. Chem. Soc. 138: 10374–10377, Journal Cover Art, 2016.

© Springer Nature Switzerland AG 2020
F. Matthäus et al. (eds.), *The Art of Theoretical Biology*, https://doi.org/10.1007/978-3-030-33471-0_57

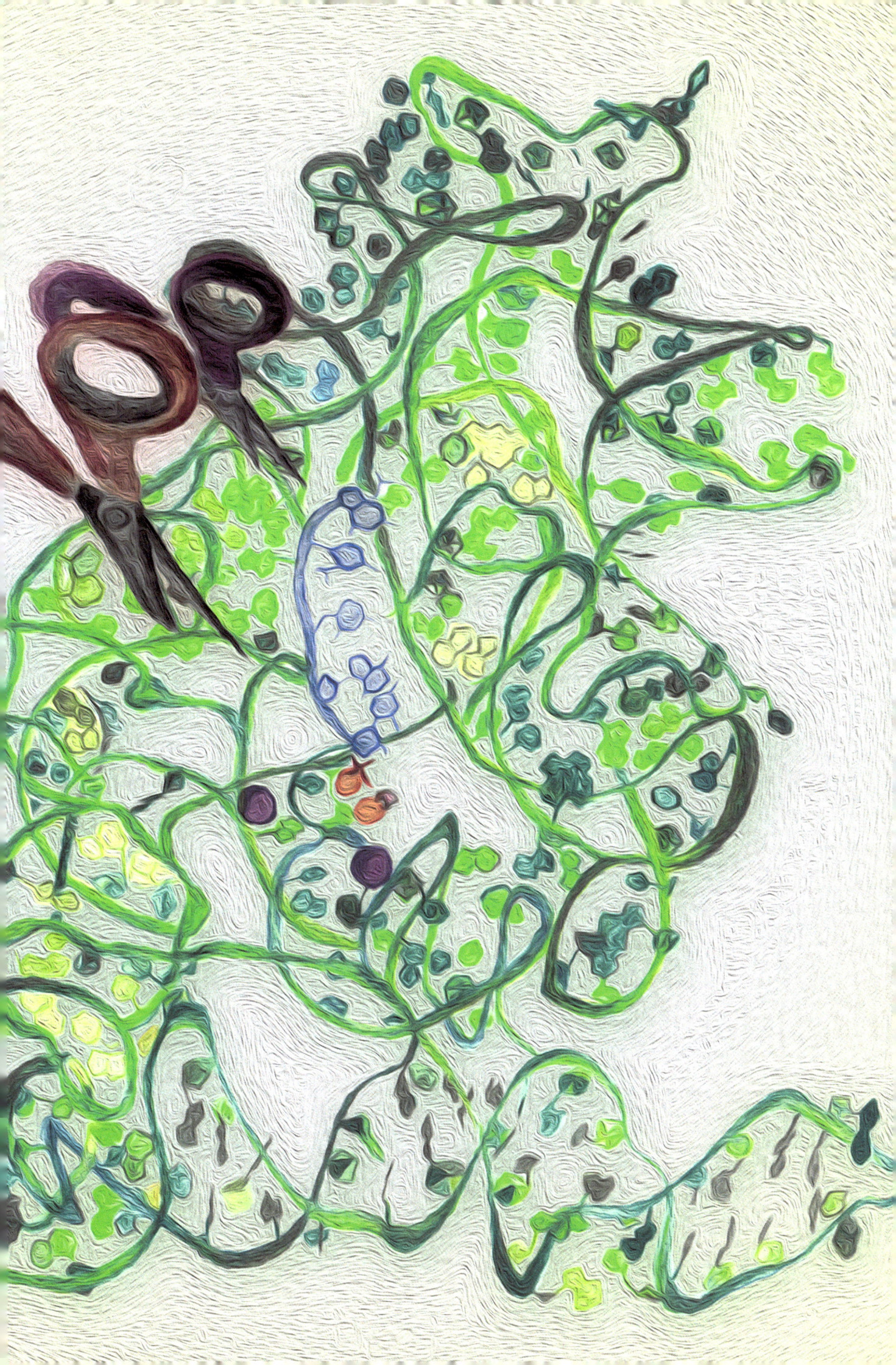

Clonal Inferno

By Ryan Schenck, Rafael Bravo & Alexander R. A. Anderson

The research story

Clonal inferno resulted from the attempt to visualize tumor growth and nutrient consumption of the growing tumors. We used a three-dimensional hybrid cellular automata model (see HAL, halloworld.org), where cancer cells are represented as round cells, while the nutrient concentration was simulated as the brown-yellow field underneath. Single tumor cells were seeded throughout the domain and grew in response to an underlying diffusible nutrient that drove cell proliferation. During division, mutations of a cells genome yielded slight differences in fitness giving rise to different shades of orange, yellows, and reds seen for the cells in the tumors. The diffusible nutrient is presented as the underlying brown-yellow rugged surface. The yellow regions underneath each tumor are areas where the nutrition has been depleted, leading to holes in the brown landscape and steep nutrition gradients. The main purpose was to investigate good representations of tumor and nutrient at the same time.

The image

The image shown was a first test, which does not quite show biological reality. We managed to find a more biological relevant representation for this problem, which produces, unfortunately, less exciting images. Hence here we show a quite interesting, but scientifically not accurate, image. The image looks a bit like growing mushrooms on a distant planet, or violent explosions. The rugged surface with long spikes underneath make for an aggressive imagery, quite fitting for cancer.

© Springer Nature Switzerland AG 2020
F. Matthäus et al. (eds.), *The Art of Theoretical Biology*, https://doi.org/10.1007/978-3-030-33471-0_58

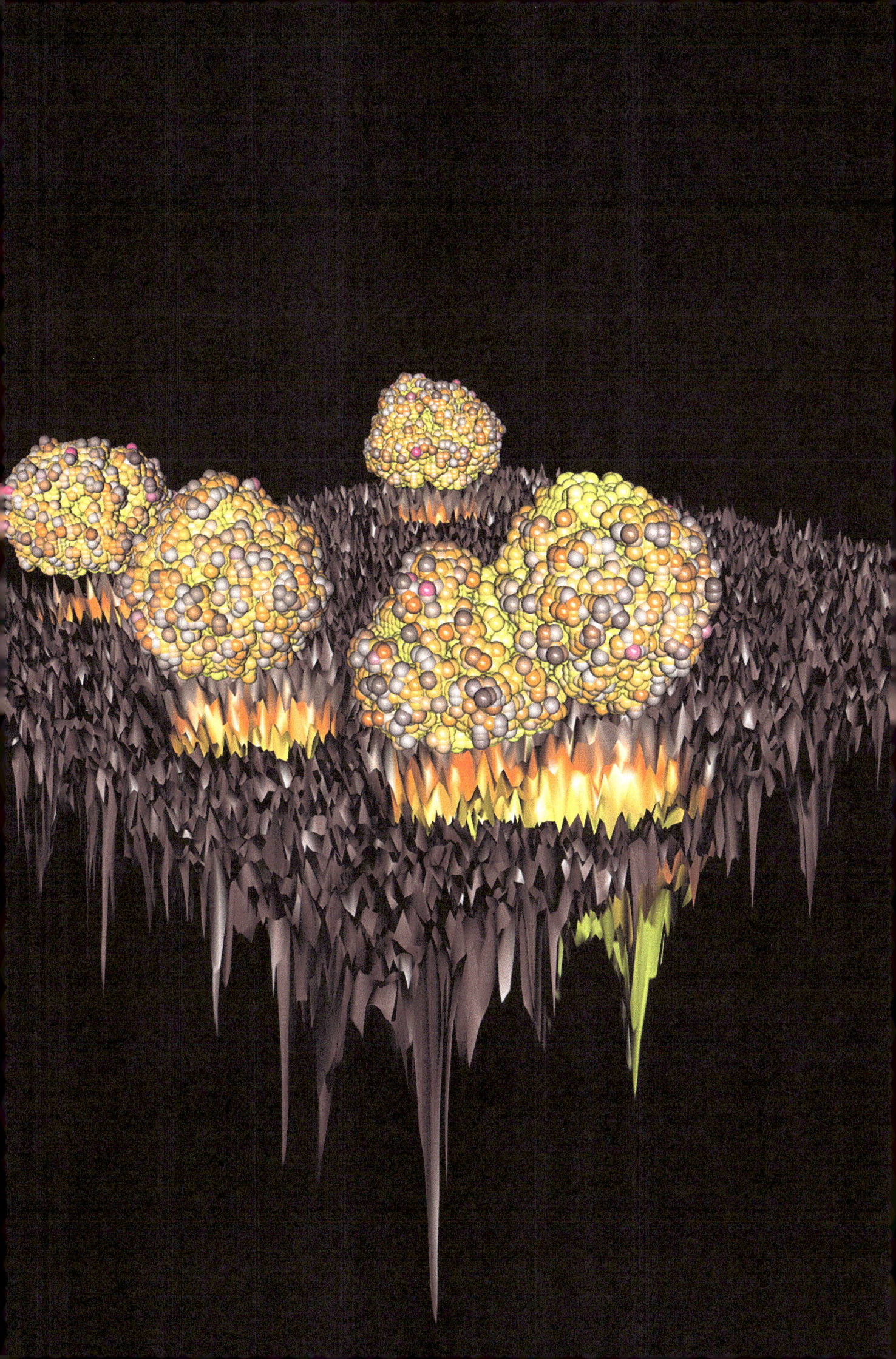

What Lies Beneath (the Heartbeat)

By Michael Colman

The research story

Understanding of the mechanisms driving the regular beating of the heart is a major challenge. The electro-mechanical system depends on multiple complex structures inside each muscle cell; because they operate over distances of just a billionth of a metre, it is difficult to dissect these structure-function relationships purely with experiments. Enter: the computer model. This study involved the development of such a model that was built using experimentally acquired imaging data, in order to simulate normal excitation in heart muscle cells. The model revealed elaborate super-resolution structure in the spatial profile of the concentration of calcium ions, the local signal for contraction.

The image

This image shows a snapshot in time of the local calcium concentration (semi-transparent colours; blue to pink indicating relative intensity) in a small portion of a simulated cell. The cell surface is visualised in the earthy-brown, with the tangled internal membrane structure just about visible as the dark-brown tubules. This frame corresponds to the peak of calcium concentration, occurring just before the maximum force; the complexity visualised persists for less than a tenth of a second. Images like this highlight just how staggeringly intricate biology can be: imagine this picture emerging in every one of the billion or so muscle cells in your heart, appearing and disappearing within a tenth of a second. Every second. Throughout your entire life. All coordinating an apparently mundane function: a simple mechanical pump, seemingly unaware of what lies beneath.

Reference

[1] Colman MA, Pinali C, Trafford AW, Zhang H, Kitmitto A, A computational model of spatio-temporal cardiac intracellular calcium handling with realistic structure and spatial flux distribution from sarcoplasmic reticulum and t-tubule reconstructions, PLoS Comput Biol. 13(8): e1005714, 2017.

© Springer Nature Switzerland AG 2020
F. Matthäus et al. (eds.), *The Art of Theoretical Biology*, https://doi.org/10.1007/978-3-030-33471-0_59

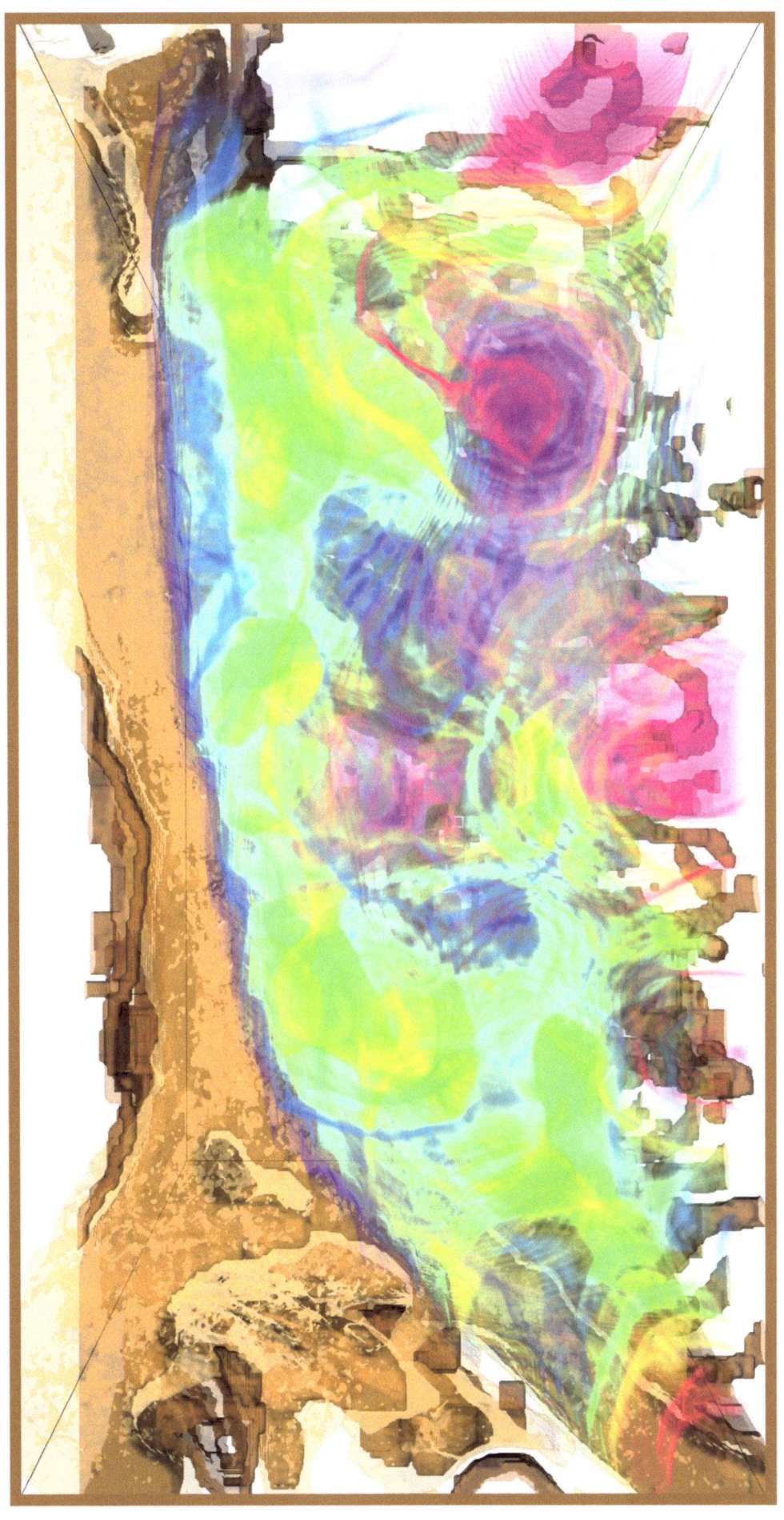

Tree of Life

By Vikram Adhikarla

The research story

A long long time ago in the era of my graduate studies, I was a poor student aspiring to simulate vasculature. I made a few cartoon simulations but nobody else liked them. I was taunted to rather stare at the trees and get some real inspiration. Pressurized for name and fame, I took the advice to heart and you have the tree up there. At that time, we were looking at patterns of vasculature and how they emerge when growing vessels experienced both directed and random cues for growth. The image represents a particular parameter set where the relative strength of directed to random cues resulted in generation of a tree-like structure. The directed cue for this growing tree was a nice and warm sun placed at the opposite end of the simulation domain. The study of vasculature was taken up to specifically study the vasculature patterns observed in the tumor and define tumor microenvironmental characteristics.

The image

The image shows a three-dimensional view of the simulated vasculature. The vasculature was simulated to have three levels with the base of the tree being the highest level. The higher two levels give the structure a direction. The capillaries form the lowest level of vessels and fill up the space between the other branches based on the required oxygen concentration. Maximum intensity projections of the image along the three axes are shown to make the image look cooler. The three levels of vessels are simulated with varying thickness and grayscale color.

References

[1] Adhikarla V, Jeraj R, An imaging-based stochastic model for simulation of tumour vasculature, Phys Med Biol 57(19): 6103–6124, 2012.

[2] Adhikarla V, dissertant, Imaging-based modeling of vasculature growth and response to anti-angiogenic therapy. Ann Arbor, MI :ProQuest LLC, 2014.

© Springer Nature Switzerland AG 2020
F. Matthäus et al. (eds.), *The Art of Theoretical Biology*, https://doi.org/10.1007/978-3-030-33471-0_60

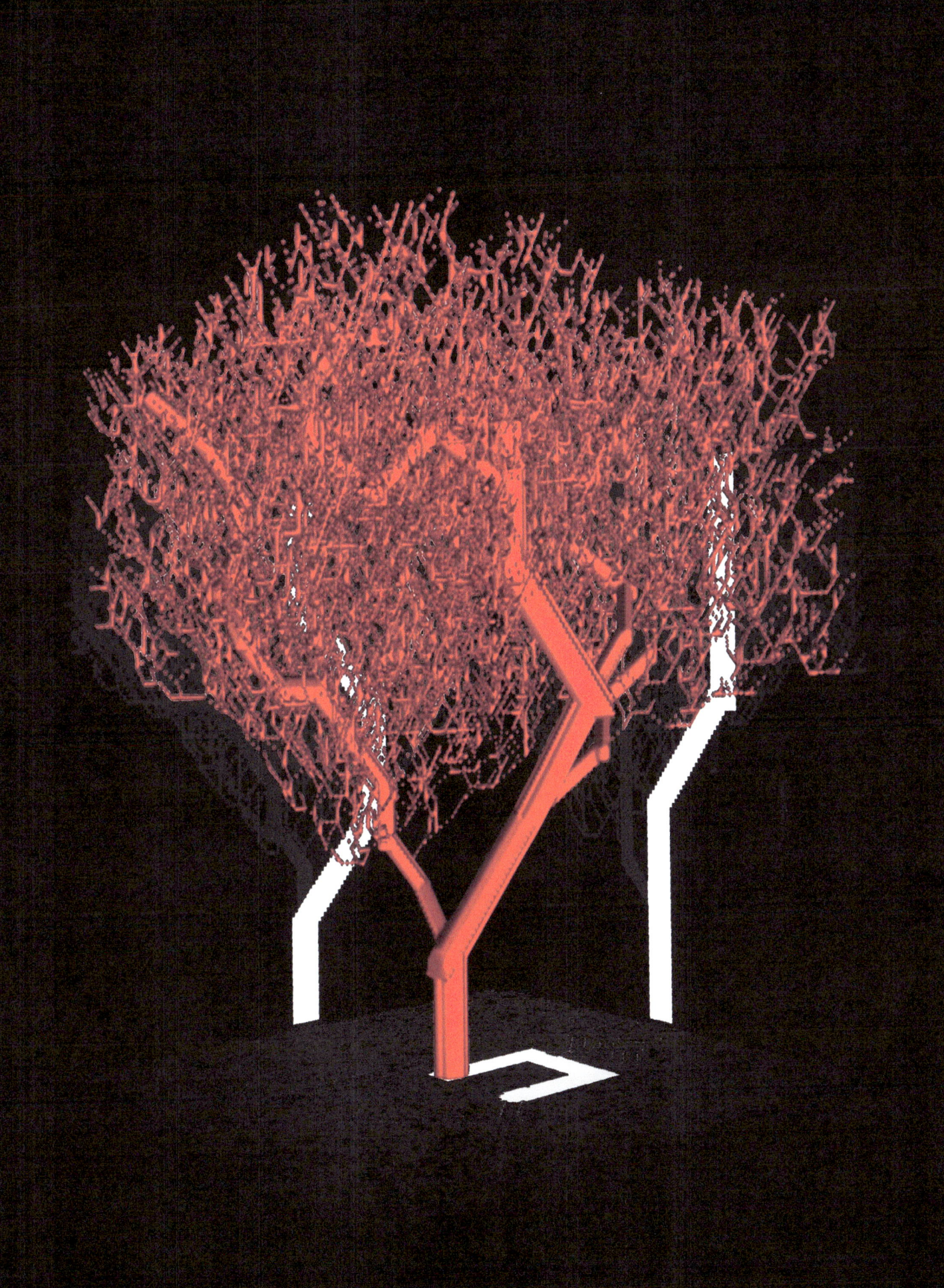

Tower of Life

By Sylvain Gretchko

The research story

Cyclic predator-prey systems are ubiquitous in nature, but little is known about how their dynamics will interact with climate change. In particular, global change is leading to higher variability in local climates. For predator-prey systems, this higher variability translates into a higher amplitude forcing of an already oscillatory system, potentially leading to highly irregular dynamics. The predator-prey system involves growth, death, and interactions of both species. Climate influence is modeled as a periodic function of time which affects the prey growth rate and describes how favourable the climate is. We investigate the effects of a climate function in which a number of "good" years is followed by an equal number of "bad" years [1]. When the climate function is constant or absent, the predator-prey system is self-oscillating. We then explore how survival of the predator is affected when we introduce the climate function at a specific phase in the predator-prey cycle.

The image

The image shows the predator survival time when we vary both the phase in the cycle when the climate function is introduced (horizontal axis) and the number of good/bad years (vertical axis). Survival time ranges from 0 (black) to the full length of the simulation (white). With a resolution of 6000 by 6000 pixels, this picture required one billion simulations. The main factor affecting survival is the phase in the cycle when the climate function is applied. The central "tower" figure corresponds to conditions where survival is most likely to happen.

Reference

[1]   Gretchko S, Marley J, Tyson RC, The effects of climate change on predator-prey dynamics, ArXiv:1805.11816 [Math, q-Bio], http://arxiv.org/abs/1805.11816, 2018.

© Springer Nature Switzerland AG 2020
F. Matthäus et al. (eds.), *The Art of Theoretical Biology*, https://doi.org/10.1007/978-3-030-33471-0_61

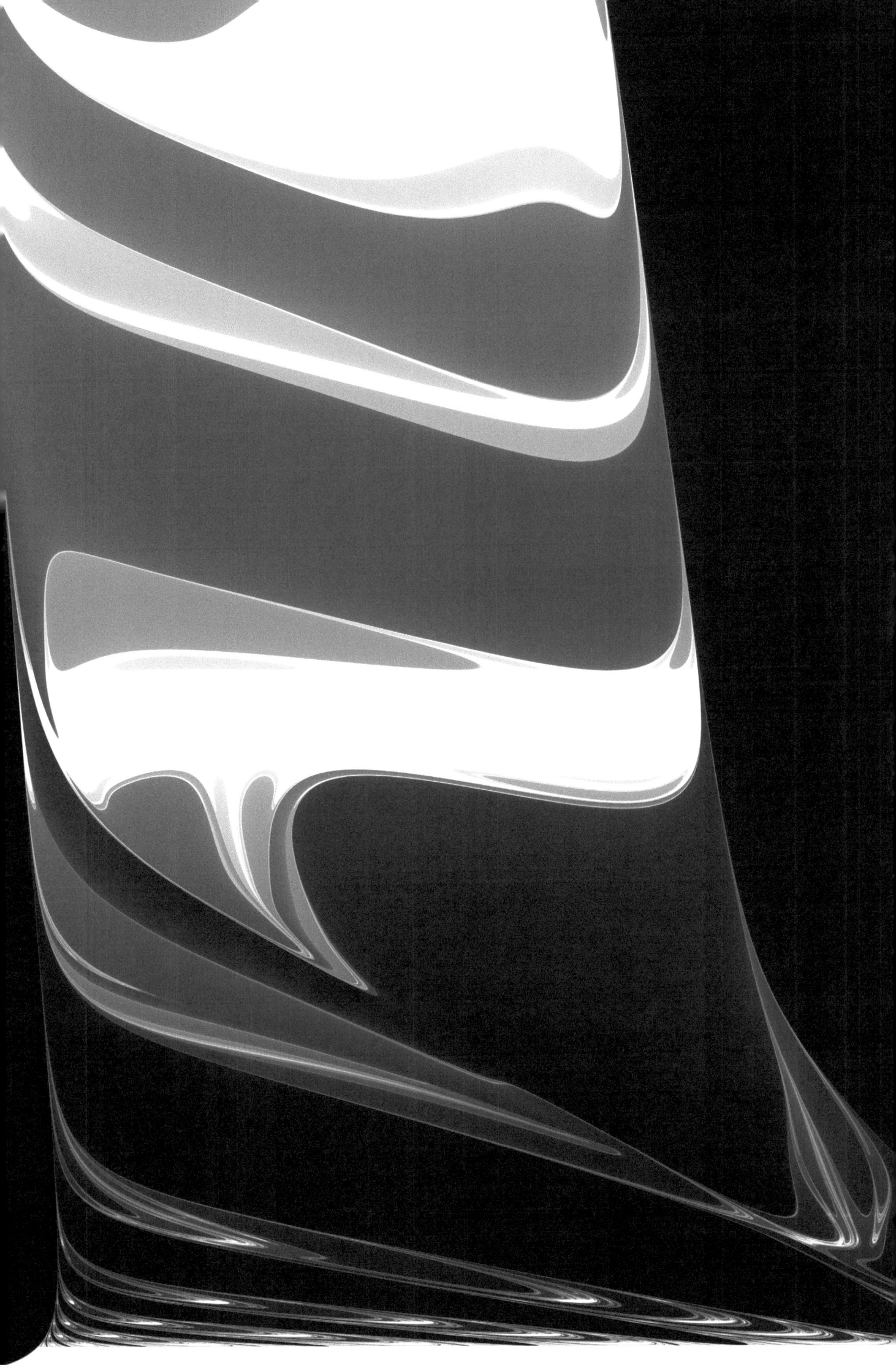

Clone Wars - The Immune System Awakens

By Maximilian Strobl, Chandler Gatenbee & Alexander R. A. Anderson

The research story

A long time ago, in a galaxy far far away (in biological distance), the ancestors of some of the cells shown as dots in this image were part of a healthy female reproductive system. But damage to their DNA caused them to proliferate and develop into an epithelial ovarian tumour. As they grew they overcame the body's defence mechanisms, enslaved the immune cells sent to kill them and recruited further cells from their surroundings. Chemotherapy provides a new hope, but only yields durable responses in a small fraction of patients. In this study we investigated the make up of ovarian tumours to understand the differences between short- and long-term responders. Novel mass-spectrometry based imaging allowed us to characterise in unprecedented detail the cancer and non-cancer cells that make up the eco-system tumour. We found that a key determinant of long-term outcome was the presence of a subset of immune cells, that once freed from their shackles, can help to reawaken the body's immune defence.

The image

Illustrated is the eco-system that is a patient's tumour. Each point represents a cell where distance between cells on the plot reflects their similarity in protein expression, and colour represents different cell types. Measurements were obtained by imaging a tumour punch biopsy with mass spectrometry (Fluidgm technology) and extracting the signatures of individual cells with an automated segmentation algorithm. This yielded the expression levels of 37 different proteins for each cell. Clusters of similar cells were identified with the DBSCAN algorithm and the data was visualised using t-SNE.

© Springer Nature Switzerland AG 2020
F. Matthäus et al. (eds.), *The Art of Theoretical Biology*, https://doi.org/10.1007/978-3-030-33471-0_62

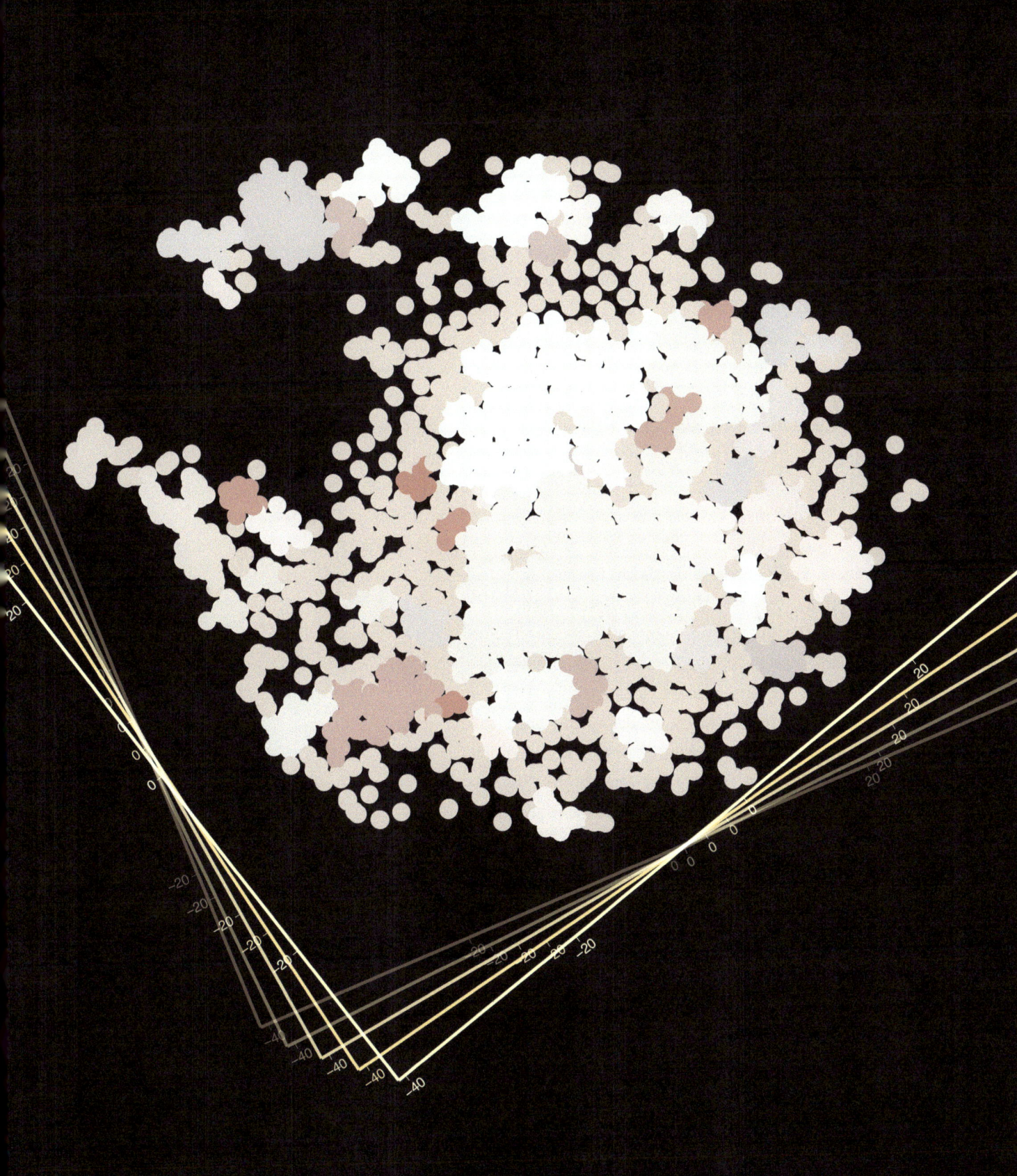

Modelled Cell

By Kai Kopfer & Franziska Matthäus

The research story

In many biological processes, such as development, healing and regeneration, or the immune response, cell migration plays an important role. The motility of cells is usually tightly controlled by chemical and mechanical cues. If this regulation fails, diseases such as cancer result. Cell motility and cell shape are regulated by the dynamic reorganization of the cytoskeleton. The cytoskeleton is built from long actin filaments, which are constantly restructured by polymerization, degradation or branching. Polymerization leads to forces protruding the leading cell front. In the rear, actin interacts with myosin to form contractile fibers. Current research focuses on the interplay between these mechanical components, actin and myosin, and small chemical compounds, called GTPases. Hereby, the chemical systems by themselves are capable of generating a symmetry break between front and back, and mechanical systems alone are able to account for various shapes that migrating cells assume. However, experimental evidence and mathematical models show that feedback exists in both ways: GTPase activity steers mechanical components, and mechanical aspects, such as tension, influence the reaction kinetics of the GTPase interaction.

The image

The cell depicted on the image has been painted using Meritum Paint, without the intention of scientific correctness. The app allows the generation of a set of extending lines by a simple finger stroke. Line style, thickness, but also the force or speed can be defined. In addition, the app uses the tablets gyroscope, and the lines bend when the tablet is moved.

© Springer Nature Switzerland AG 2020
F. Matthäus et al. (eds.), *The Art of Theoretical Biology*, https://doi.org/10.1007/978-3-030-33471-0_63

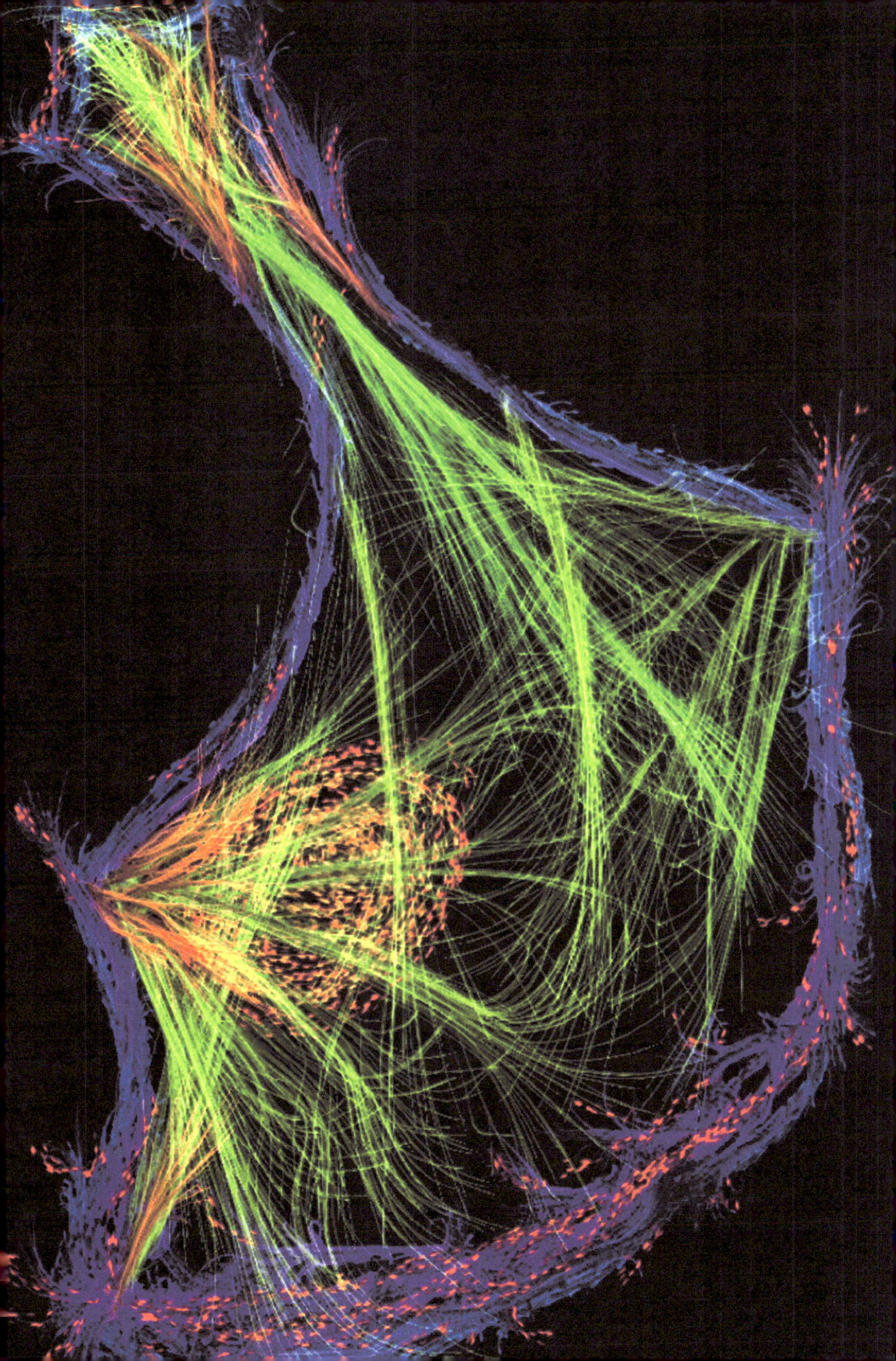

CD196-

By Chandler Gatenbee, Alexander R. A. Anderson & Maximilian Strobl

The research story

Cancer cells do not exist in isolation, but are part of a larger ecology that contains other cell types and resources. Understanding the tumor ecology can help discern which environments allow tumors to thrive and which cause them to recede. The tumor ecology can be described quantitatively using data collected from imaging techniques that identify cell types in the tumor based on the proteins they express. By marking each cell type and knowing their location in space, one can infer how the various cell types are interacting. Combined with patient outcome data such as response to therapy, it is possible to associate ecology with outcomes and generate hypotheses about how ecology shapes tumor evolution through progression and in response to treatment.

The image

The image here is derived from a circular biopsy taken from a patient with ovarian cancer. The location of cell nuclei and expression levels of 38 proteins were measured in the sample. Using computer vision techniques, the boundaries of each cell were estimated based on the distance between nuclei. In the presented image, each cell is colored according to whether or not it expresses CD196, a protein found primarily on a subset of immune cells: grayscale cells express CD196, while colored cells do not express CD196 (i.e. they are CD196-). The large triangular tiles denote areas outside of the biopsy. This process can be repeated for each marker, creating a cell type map that describes the tumor ecology.

© Springer Nature Switzerland AG 2020
F. Matthäus et al. (eds.), *The Art of Theoretical Biology*, https://doi.org/10.1007/978-3-030-33471-0_64

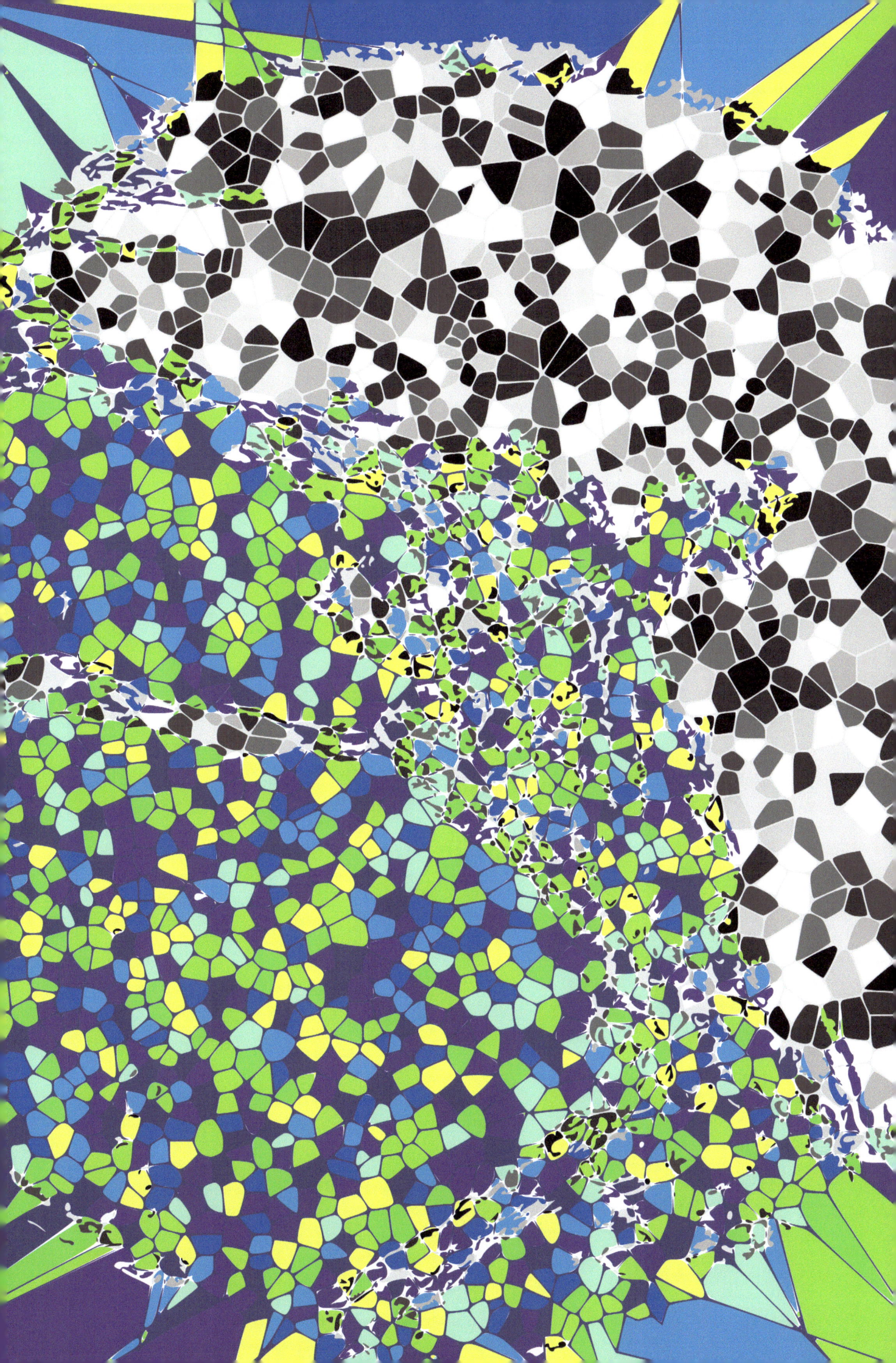

Tumor Composition Depends on the Viewing Angle

By Jan Poleszczuk & Heiko Enderling

The research story

In a typical setting, mathematical biologists develop models with a certain number of free parameters, either to be calibrated with experimental data, or to be evaluated numerically to explore a variety of model behaviors. Back in 2014 we decided to investigate a tumor growth model in which each simulated cell can have its own set of parameters (or traits) that may randomly change in time, mimicking acquired mutations [1]. Our goal was to evaluate how cancer growth patterns are affected by the evolutionary trajectories of random mutations, and which cellular traits are most important for aggressive growth.

The image

We performed hundreds of independent model simulations and selected the tumor that grew the biggest in the given timespan. In the center of the presented image we can depict the simulated tumor composition, whereby each pixel represents a single cell with its color corresponding to the individual trait, i.e. darker colors represent more aggressive cell behaviors. A circular histogram surrounding the tumor shows the results of an in silico biopsy experiment of expected trait heterogeneity. To this extent we have collected on average 100,000 cells at different angles in 5 degrees intervals and calculated the mean and standard deviation of a selected cellular trait. We can see that the derived information for any trait can vary orders of magnitude dependent on biopsy angle.

Reference

[1] Poleszczuk J, Hahnfeldt P, Enderling H, Evolution and phenotypic selection of cancer stem cells, PLoS Computational Biology 11(3): e1004025, 2015.

© Springer Nature Switzerland AG 2020
F. Matthäus et al. (eds.), *The Art of Theoretical Biology*, https://doi.org/10.1007/978-3-030-33471-0_65

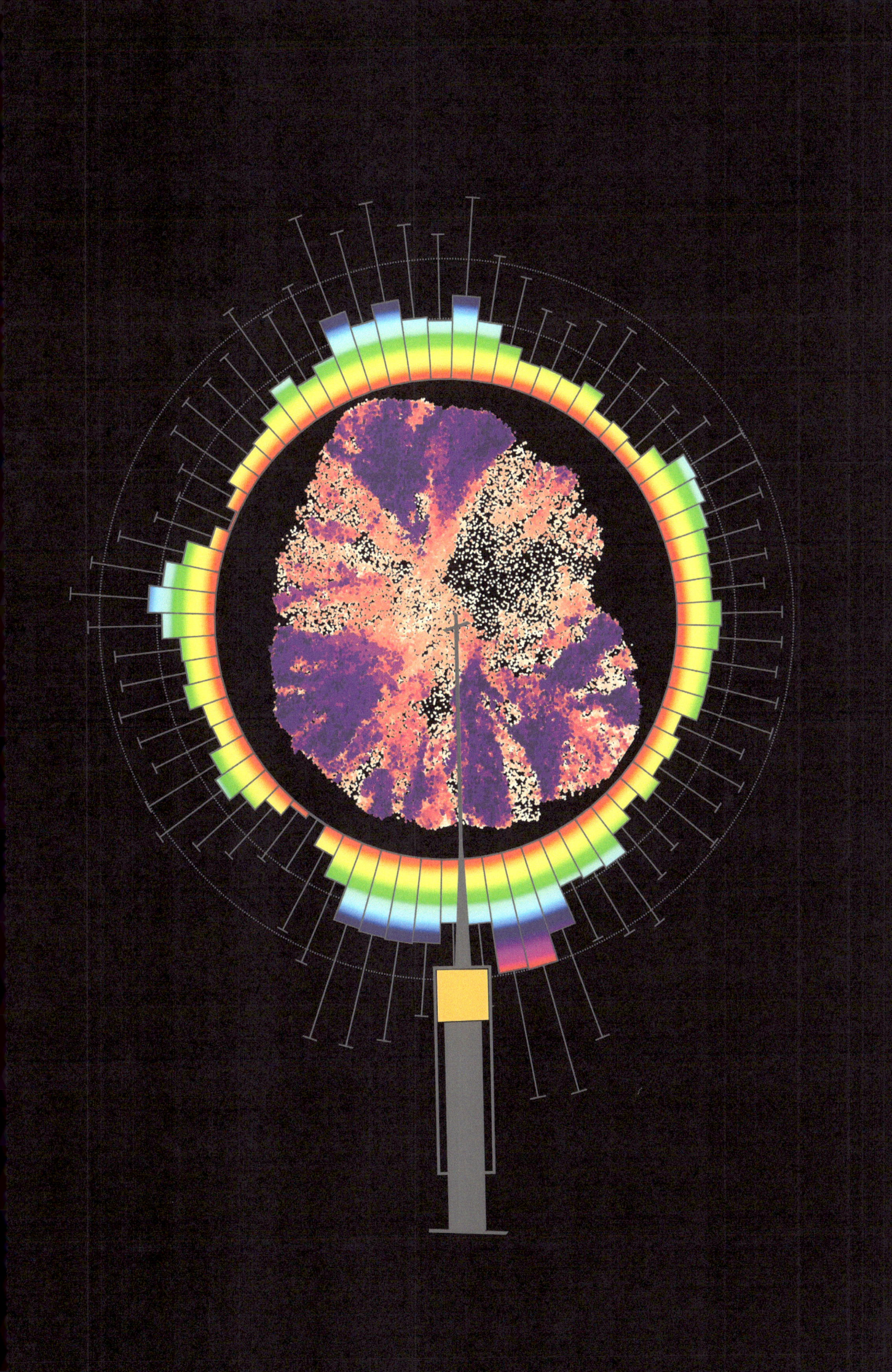

Poincaré's Homoclinic Horror

By Stefanos Folias

The research story

A simplified mathematical model, in the form of a discrete dynamical system, was developed to study the spatially-localized, synchronous spiking observed in a large network of neural tissue. The model involved populations of excitatory and inhibitory neurons, interacting via synapses, and driven by a localized current input. Given that activity patterns of large populations of coupled neurons are notoriously difficult to study analytically, the motivation of this project was to develop an analytically-tractable approach to describe the spatial structure of the synchronous oscillations as the system parameters are varied. The simplified mathematical model is an implicit discrete map with two variables describing, on each cycle of a gamma frequency oscillation (30 – 90 Hz), the spatial extent of a synchronous spiking activity in the excitatory and inhibitory neuronal populations, respectively.

The image

The image visualizes a solution to the discrete map as ordered pairs in the phase plane, generated from 1 million iterations of the map. A chaotic solution occurs in the presence of a double homoclinic tangle, an extraordinarily intricate structure, arising in a 1:2 resonance bifurcation (codimension 2 bifurcation) as parameters are varied. Henri Poincaré first described the homoclinic tangle in 1899, after a glimpse in 1890, in the context of the famous three-body problem. Without the aid of computers to visualize such complicated solutions, he understood – from the mathematics alone – the implications such a structure would have on solutions and their characterization, referring to it in "horror." He, in fact, had discovered a primary mechanism that gives rise to chaos.

Reference

[1] Folias SE, Ermentrout GB, Spatially localized synchronous oscillations in synaptically coupled neuronal networks: conductance-based models and discrete maps. SIAM J. Applied Dynamical Systems 9: 1019–1060, 2010.

© Springer Nature Switzerland AG 2020
F. Matthäus et al. (eds.), *The Art of Theoretical Biology*, https://doi.org/10.1007/978-3-030-33471-0_66

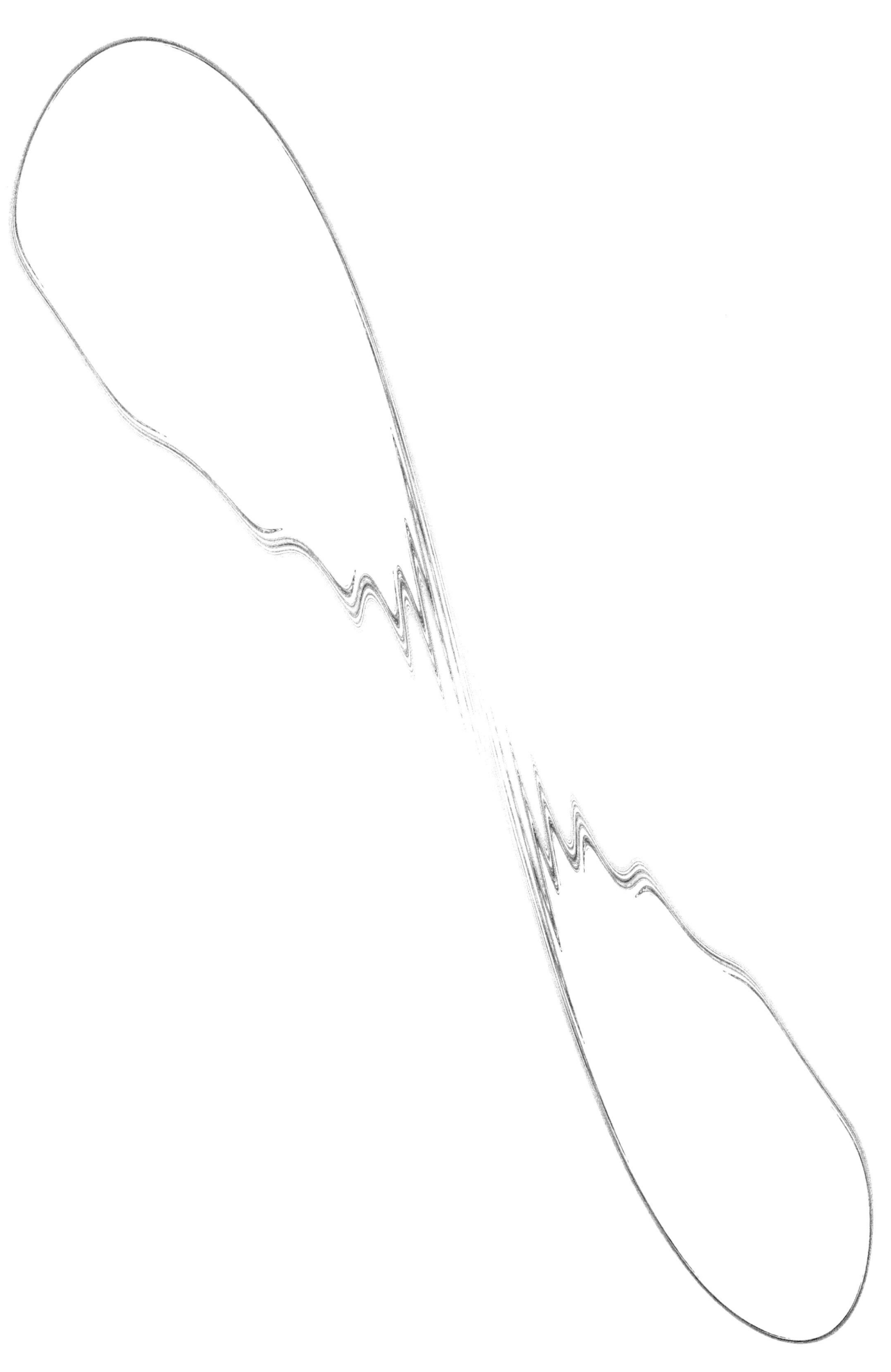

E|A|S (Evolving Asteroid Starships)

By Nils Faber & Angelo C.J. Vermeulen

The research story

E|A|S (Evolving Asteroid Starships) is a trans-disciplinary research project in which bio-inspired concepts for (manned) interstellar exploration are being developed. A long-duration journey through interstellar space is characterized by a high level of uncertainty [1]. Environmental disturbances such as cosmic radiation surges and particle impact events cannot be predicted in detail for the entire flightpath. A spacecraft with a built-in capacity to grow and evolve during its journey offers a solution to cope with such unforeseen challenges.

The image

The spacecraft concept shown in this artist's impression relies on asteroid mining and onboard 3D manufacturing. An asteroid is redirected and gradually transformed into a hybrid spacecraft. Asteroid mining provides the resources for ongoing 3D manufacturing of the spacecraft's architecture. Using a morphogenetic engineering approach [2], the spacecraft develops itself gradually, both inside and outside the asteroid. The modular nature of the spacecraft enables structural and functional reconfiguration of its architecture. This allows for an ongoing morphological evolution to adapt and cope with unexpected environmental changes. The E|A|S project focuses on creating a hybrid computer simulation in which this morphogenetic engineering approach to interstellar exploration can be tested. The images are based on two 3D models created in Blender and respectively show an overview of the proposed starship concept and a cutaway view.

References

[1] Klessen RS, Glover SCO, Physical processes in the interstellar medium, arXiv: 1412.5182v1 [astro-ph.GA].

[2] Doursat R, Sayama H, Michel O (Eds.), Morphogenetic Engineering, Springer, Heidelberg, 2012.

© Springer Nature Switzerland AG 2020
F. Matthäus et al. (eds.), *The Art of Theoretical Biology*, https://doi.org/10.1007/978-3-030-33471-0_67

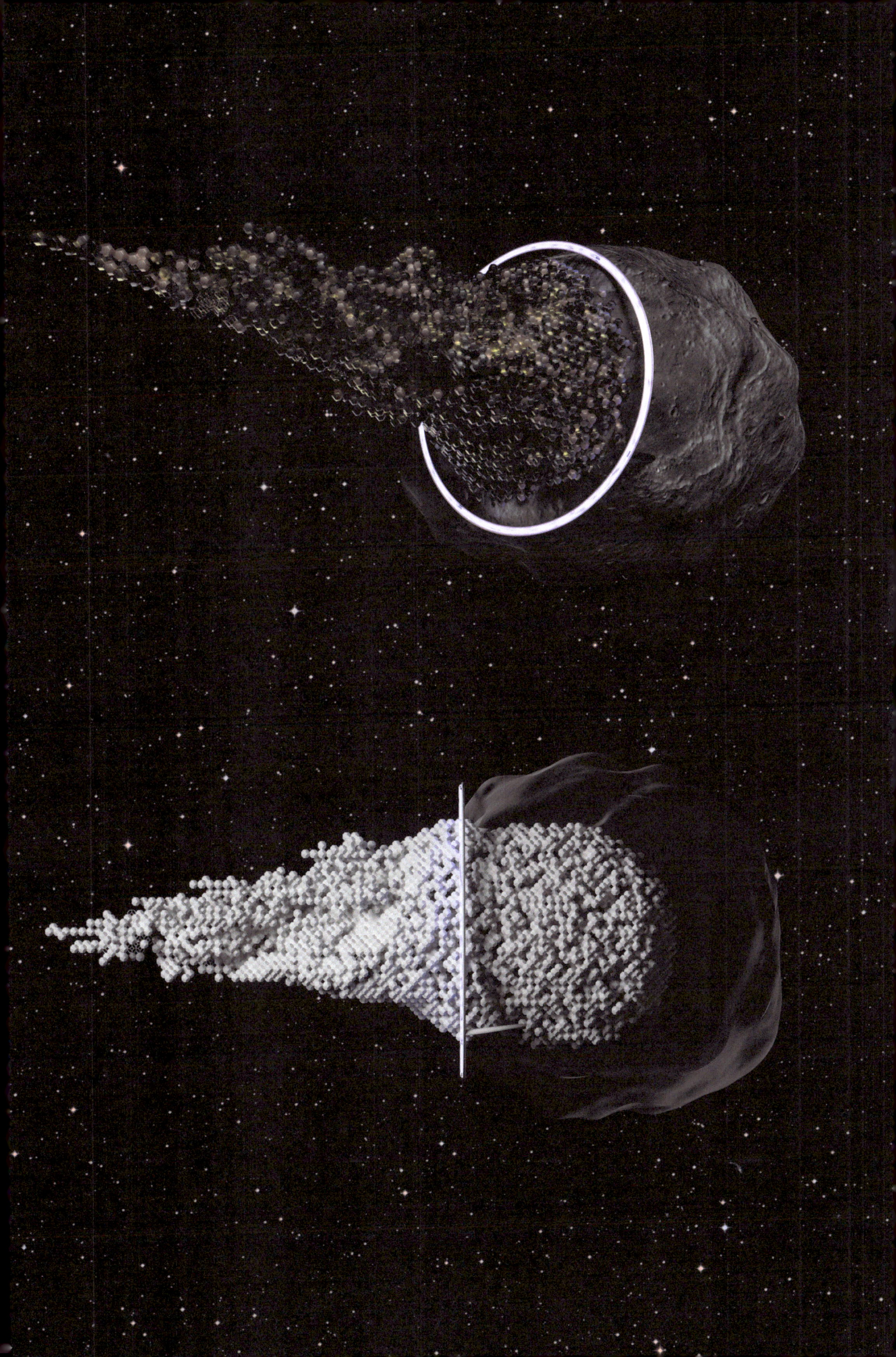

Interacting Spider Webs

By Jens Rieser, Daniel Bruneß, Jörg Ackermann & Ina Koch

The research story

The bacterial pathogen *Salmonella* Typhimurium provokes gastroenteritis and typhoid fever. *Salmonella* become multidrug resistant. The infection by *Salmonella* affects the dynamics of regulation events, such as phosphorylation of the host proteins. The understanding of these dynamics can help to find possible drug targets to support the cellular immune response and to prevent the pathogenic effects of the infection [1]. We wanted to investigate the protein-protein interaction network (PIN) of *Salmonella*-infected HeLa cells. To identify the proteins, scientists have analyzed the infected cells by mass spectrometry methods [2]. Knowing the proteins, we built the PIN. These networks consist of vertices, which represent the proteins and edges that represent an experimentally observed interaction between two proteins.

The image

The data for the interaction are publicly available in various online databases such as STRING, BioGRID or IntAct. In this image, the PIN includes 1034 phosphorylated proteins and 5986 interactions. The color of the edges represents the edge betweenness as a measurement of communication in the network. The coloring ranges from green to red depending on how many proteins in the network may communicate through the corresponding edge. Different clustering methods led to this particular segregation of the proteins, which exhibits special relationships among the proteins in one cluster, such as functional or structural similarities, dependencies and membership of protein families or protein complexes. Using these clustering methods and data integration, the analysis of several network properties can reveal interesting insights, such as driver proteins and bottlenecks in the signaling pathways of infected organisms.

References

[1] Pan A, Lahiri C, Rajendiran A, Shanmugham B, Computational analysis of protein interaction networks for infectious diseases, Briefings in Bioinformatics, 17.3:517–526, 2015.

[2] Rogers LD, Brown NF, Fang Y, Pelech S, Foster LJ, Phosphoproteomic analysis of Salmonella-infected cells identifies key kinase regulators and SopB-dependent host phosphorylation events, Science Signal 4:191, 2011.

© Springer Nature Switzerland AG 2020
F. Matthäus et al. (eds.), *The Art of Theoretical Biology*, https://doi.org/10.1007/978-3-030-33471-0_68

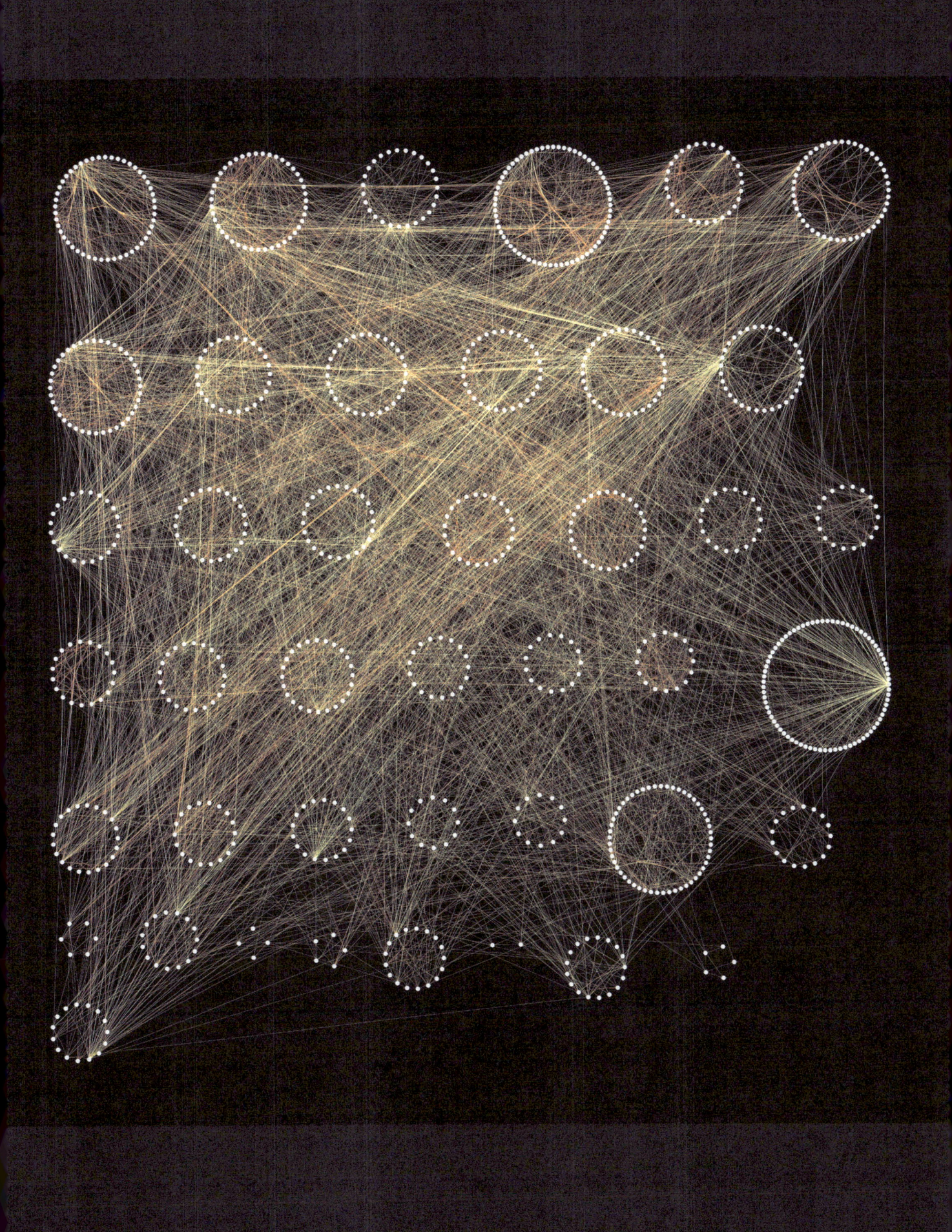

Heart Cells are aMAZEing

By Nina Kudryashova & Alexander Panfilov

The research story

The main function of the heart is pumping blood, which is performed by contraction of the cardiac cells (red). However, there are also other cell types composing the heart, including a large fraction of fibroblasts (cyan): connective cells which glue the tissue together and repair damaged regions. The number of fibroblasts gradually increases with age. The problem is that cardiac cells use electrical signals to synchronize their contraction, but fibroblasts block these. In order to overcome it, cardiac cells organize themselves into a MAZE of conducting pathways and in such way that even a small number of cardiac cells can still remain connected and transmit electrical signals. This maze was first discovered by my colleagues in Moscow in cell culture experiments. However, the mechanism behind such network formation was not clear. Using a novel model of a virtual cardiac tissue, which combines ideas of cardiac modelling and developmental biology [1,2], we were able to identify these mechanisms and reproduce such cardiac maze in a computer simulation.

The image

This picture shows a maze organized by cardiac cells (red), which we think is required to fight against ageing and damage. In this image, there is only 30% of red cardiac cells, but they successfully wire the whole sample. Therefore, our study shows that heart cells have aMAZEing capacity for maintaining their main function: collective synchronous contraction, which they accomplish by their ability to build a MAZE.

References

[1]   Kudryashova N, Nizamieva A, Tsvelaya V, Panfilov AV, Agladze KI (2019) Self-organization of conducting pathways explains electrical wave propagation in cardiac tissues with high fraction of non-conducting cells. PLoS Comput Biol 15(3): e1006597.

[2]   Kudryashova N, Tsvelaya V, Agladze K, Panfilov A, Virtual cardiac monolayers for electrical wave propagation, Scientific Reports 7(1): 7887, 2017.

© Springer Nature Switzerland AG 2020
F. Matthäus et al. (eds.), *The Art of Theoretical Biology*, https://doi.org/10.1007/978-3-030-33471-0_69

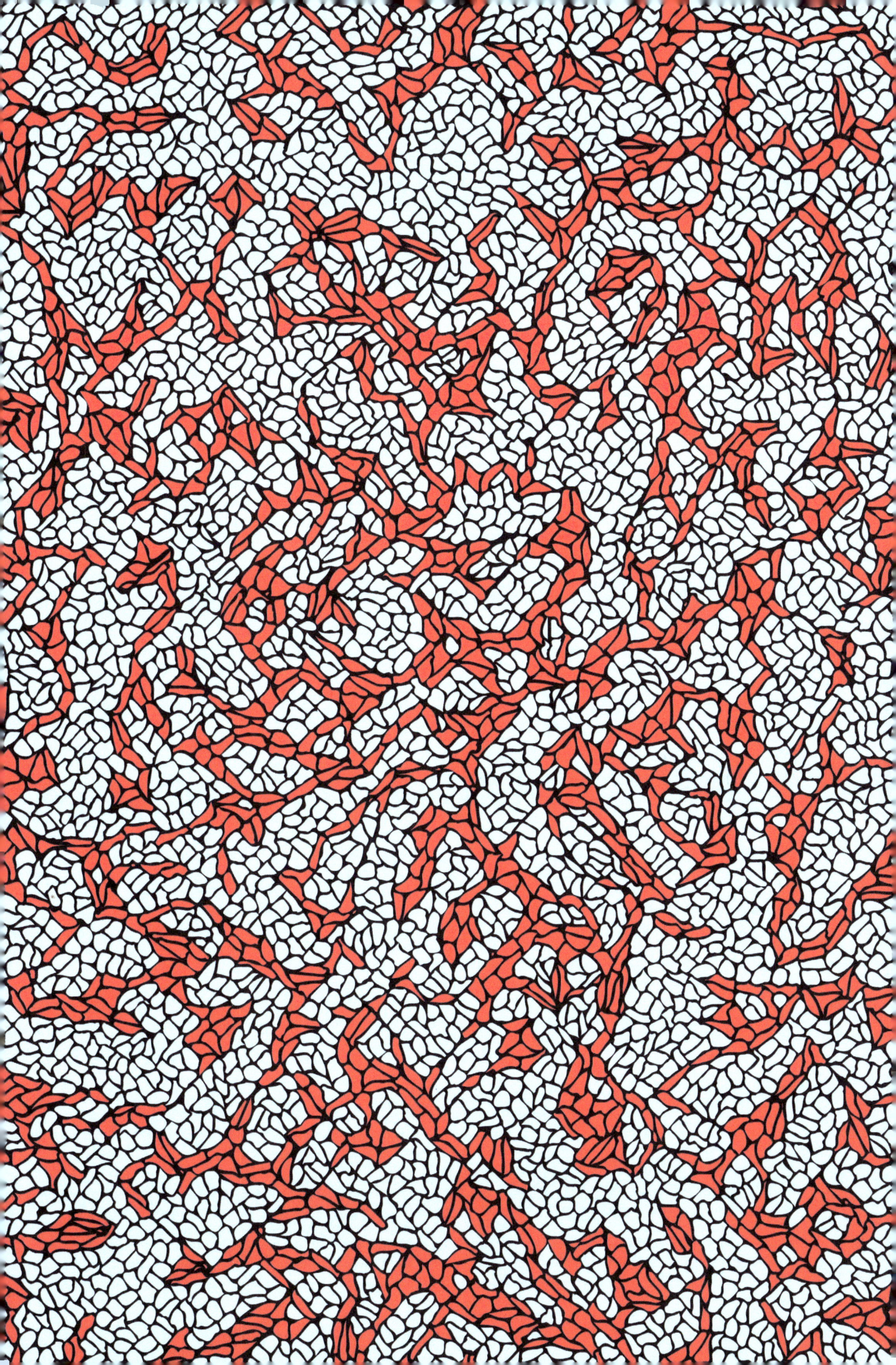

Actin Spring

By Gaëlle Letort

The research story

Actin filaments are major actors of cell division, migration, morphological changes and other cellular processes. These cytoskeletal filaments, short, very dynamic and flexible, but numerous, assemble into higher level structures that contribute to those different processes. The specific organization of thousands of actin filaments in those different structures determine their ability to perform different tasks. Understanding how from the same biological entities, cells can build, use and disassemble such different structures in form and function is thus a major point in understanding cellular processes. To explore the auto-organization of actin filaments, controlled in-vitro reconstructions were performed by constraining actin filaments generation to given geometries. Filaments will grow and assemble reproducibly into structures determined by the imposed geometry [1]. However, despite the in-vitro control, the complexity of the system makes it difficult to explain the observed behavior and to establish the laws controlling filaments organization. To gain understanding of this emergent organization, we performed numerical simulations of such systems with the Cytosim software.

The image

This image is the result of one simulation of the auto-organization of actin filaments growing out of an 8 branch star geometry, similar as those performed in [2]. The image was directly generated from Cytosim and was converted to eps format with Gimp.

References

[1] Reymann AC, Martiel JL, Cambier T, Blanchoin L, Boujemaa-Paterski R, Théry M, Nucleation geometry governs ordered actin networks structures, Nature Materials 9(10): 827–832, 2010.

[2] Letort G, Politi AZ, Ennomani H, Théry M, Nédélec F, et al., Geometrical and mechanical properties control actin filament organization, PLoS Computational Biology 11(5): e1004245, 2015.

© Springer Nature Switzerland AG 2020
F. Matthäus et al. (eds.), *The Art of Theoretical Biology*, https://doi.org/10.1007/978-3-030-33471-0_70

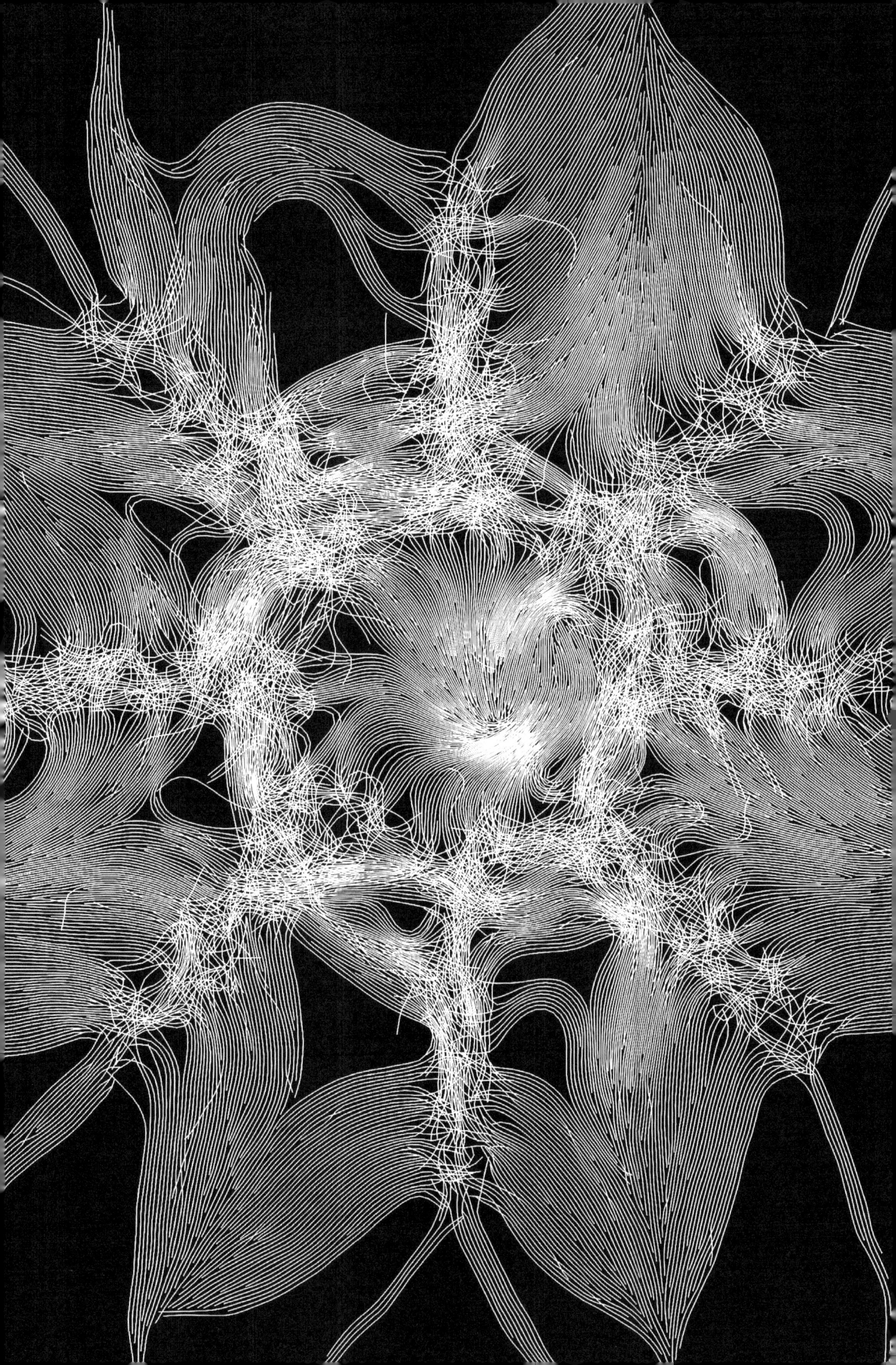

To model animal abundance patterns at ea
positional probability density function u
satisfies the advection diffusion equa

$$\frac{\partial u}{\partial t}(x,t) = \frac{\tau}{2}\frac{\partial^2}{\partial x^2}\left(D(x)u(x,t)\right) - \frac{\partial}{\partial x}$$

with

$$c(x = \begin{cases} \frac{\sigma}{\tau\sqrt{2\pi}\,g(x)}\left[(\beta-1)\exp\left(\frac{-1}{}\right.\right. \\ \frac{\sigma}{\tau} \quad \left[\quad \right)\exp\left(\frac{-1}{}\right. \end{cases}$$

$$\exp\left(\frac{-x^2}{2\sigma^2}\right) -$$

$$\times\sqrt{\frac{\pi}{2}}\,\text{erf}\left(\frac{x}{\sqrt{2}\sigma}\right.$$

$$\left(\frac{x}{\sqrt{2}\sigma}\right) + \text{erf}$$

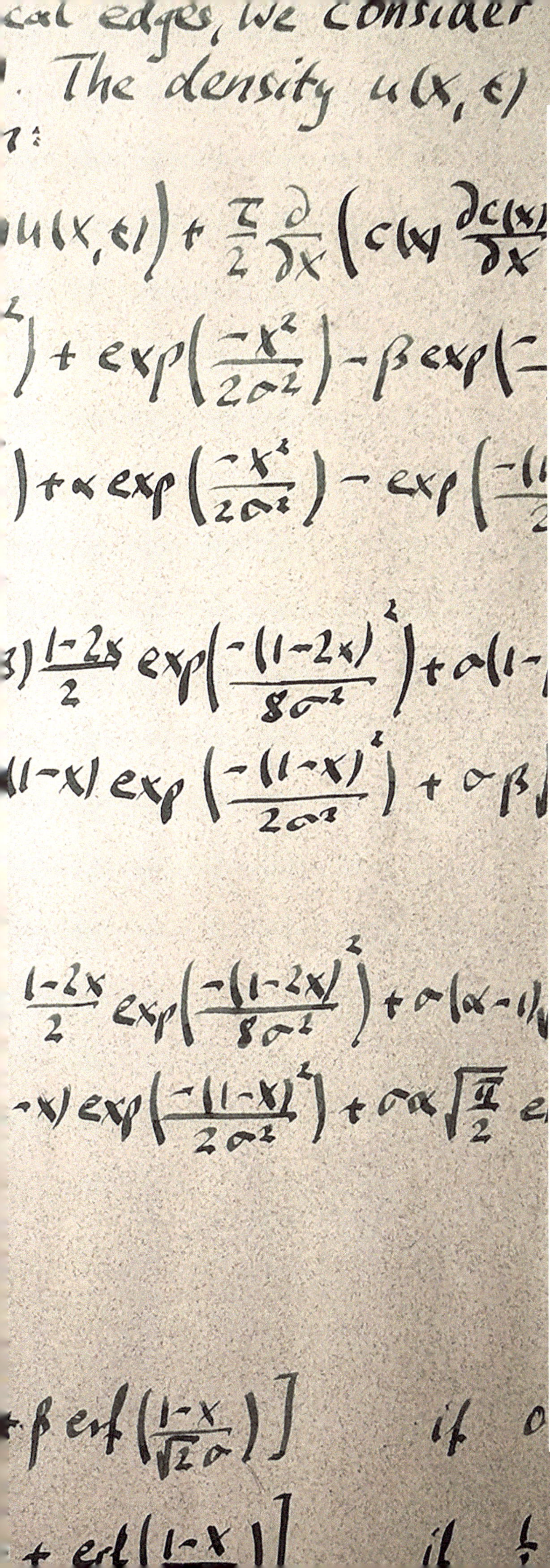

Handwritten background (partially visible):

cal edges, we consider the

The density $u(x, \epsilon)$

$u(x, \epsilon) + \frac{\tau}{2}\frac{\partial}{\partial x}\left(c(x)\frac{\partial c(x)}{\partial x}\right)$

$+ \exp\left(\frac{-x^2}{2\sigma^2}\right) - \beta \exp\left(-\right.$

$+ \alpha \exp\left(\frac{-x^2}{2\sigma^2}\right) - \exp\left(\frac{-1}{2}\right.$

$\frac{1-2x}{2}\exp\left(\frac{-(1-2x)^2}{8\sigma^2}\right) + \alpha(1-$

$(1-x)\exp\left(\frac{-(1-x)^2}{2\sigma^2}\right) + \sigma\beta$

$\frac{1-2x}{2}\exp\left(\frac{-(1-2x)^2}{8\sigma^2}\right) + \sigma(x-1)$

$-x)\exp\left(\frac{-(1-x)^2}{2\sigma^2}\right) + \sigma\alpha\sqrt{\frac{\pi}{2}}e$

$+\beta \, erf\left(\frac{1-x}{\sqrt{2}\sigma}\right)\Big]$ if   o

$+ \, erf\left(1-x\right)\Big]$ if $\frac{1}{2} \leq x \leq 1$.

2017

Dynamical Diggers

By Jessa Marley, Thomas Hillen,
& Nick Parayko

The research story

This work was created to explore the connection between the mathematical models and the real-world inspiration behind it. The effect of edges of an environment occurs in both individual animals and animal populations. For example, the effect of bordering field types on the distribution of Richardson's ground squirrels. The formula in the background of the art is an explicit solution of a species distribution along an edge in the environment [1]. The partial differential equations could be applied to not only rodent movement but other animals as well. These equations were made to understand how animals respond to the edge of two conjoining habitat types and link together the behaviour of the individual to that of the population. The edge effects that emerge depend on model parameters, and the distance to the edge where the animals make their decisions is key to determine the magnitude of the effects.

The image

The photographic inspiration is of a mother and baby Richardson's ground squirrels surveying their surroundings. The background mathematics was made using a calligraphy pen and green ink. The ground squirrels and floral were drawn using chalk pastels. With the two components together we aim to give a visualization for the connection behind mathematical biology research and the real world.

Reference

[1]   Potts JR, Hillen T, Lewis MA, Edge effects and the spatio-temporal scale of animal movement decisions, Theoretical Ecology 9(2): 233–247, 2015.

Author Information

Daniel Abler is a postdoctoral research fellow in mathematical oncology and computational biomechanics at the City of Hope National Medical Center, Duarte, California. He is interested in modeling and simulation of biomedical processes, often based on (and producing) imaging data.

Contributed image: page 14

Jörg Ackermann is a Senior Scientist in the Molecular Bioinformatics Group of the Goethe University Frankfurt, Germany. His interests are networks, dynamical systems, stochastic processes, and development of algorithms for proteomics, metabolism, and signaling pathways.

Contributed images: pages 112 and 136

Vikram Adhikarla is a Medical Physicist and a biomathematical modeler at the City of Hope National Medical Center, Duarte, California. His primary interests are trees and dragons. The rest of life follows.

Contributed images: pages 14 and 120

Ishan Ajmera is a Research Fellow at the Centre for Plant Integrative Biology (CPIB), University of Nottingham, UK, in the area of computational systems biology. His research interests lie in using mathematical and computational techniques as tools to address biological questions.

Contributed images: pages 22 and 66

Alexander R.A. Anderson is Chair of the Mathematical Oncology Department at the Moffitt Cancer Center, in sunny Tampa Bay, Florida. He is interested in using mathematics to develop better cancer treatments that are less toxic and longer lasting. He cares deeply about sharing science clearly, freely, and visually.

Contributed images: pages 60, 70, 76, 116, 124 and 128

Chaitanya A. Athale is Associate Professor at the Indian Institute of Science Education and Research (IISER), Pune, India. He heads the group Self-Organization and Cell Morphogenesis since 2009. He is interested in understanding biological spatial patterns through experiments and simulations.

Contributed image: page 86

Daniele Avitabile is Associate Professor of applied mathematics in the Department of Mathematics at Vrije Universiteit Amsterdam. He is interested in pattern formation in infinite-dimensional dynamical systems, mathematical modelling of biological processes, and computational applied mathematics.

Contributed image: page 32

Francisco Azuaje leads the Bioinformatics and Modeling Research Group (BIOMOD) at the Luxembourg Institute of Health (LIH). He also enjoys nature, literature, history, films, learning languages, and cooking.

Contributed image: page 72

Ruth Baker is Professor of applied mathematics within the Mathematical Institute at Oxford University. Her research interests lie in developing and using mathematical and computational approaches to explore key problems in cell and developmental biology.

Contributed image: page 48

Leah Band is an Assistant Professor at the University of Nottingham. She is a mathematician and develops mathematical and computational models to understand how plants grow and develop. She works closely with plant scientists to develop and test her model predictions.

Contributed images: pages 22 and 66

Michael Barish is a Professor of neurobiology at the City of Hope National Medical Center, Duarte, California, with an interest in anything strange and curious. His present interest is in understanding heterogeneity of glioblastoma and its relevance to developing and refining immunotherapies.

Contributed image: page 98

Daniel Barker is Reader in bioinformatics at the University of Edinburgh. He is particularly interested in phylogeny reconstruction, bioinformatics education, and philosophy of science.

Contributed image: page 6

Nicholas Battista is Assistant Professor of mathematics at the College of New Jersey, in Ewing Township, NJ, USA. He is passionate about biomechanics, studying how things move and work, and sharing his love of science (aka asking questions) with others.

Contributed image: page 106

Katharina Baum works as postdoctoral researcher at the Luxembourg Institute of Health (LIH), Luxembourg, and the Max Delbrück Center for Molecular Medicine in the Helmholtz Association (MDC). She is fascinated by solving biological riddles with mathematical tools, especially revealing mechanisms and functions responsible for complex diseases such as cancer.

Contributed image: page 72

© Springer Nature Switzerland AG 2020
F. Matthäus et al. (eds.), *The Art of Theoretical Biology*, https://doi.org/10.1007/978-3-030-33471-0

Katharina Becker is a Master's student at the University of Heidelberg. She is fascinated by both, order and chaos, and likes creating one from the other.
Contributed image: page 8

Marek Bodnar is member of the Section of Biomathematics and Game Theory at the Faculty of Mathematics, Informatics and Mechanics, University of Warsaw. His main scientific interests are focused on delay ordinary differential equations applied to models describing biological phenomena.
Contributed image: page 62

Rafael Bravo is a Research Associate at the Moffitt Cancer Center, Tampa Bay, Florida. His focus is on designing efficient agent based models for biological research and on creating a generalized framework to accelerate model development and unify several projects. He plans to use his modeling skills to study cancer.
Contributed image: page 116

Kai Breuhahn is a Research Fellow at the Institute of Pathology, University Hospital Heidelberg, Germany, focussing on the mechanisms of cancer development and tumor progression. One key aspect of his work is to understand how tumor cells acquire their ability to move and disseminate.
Contributed image: page 26

Daniel Bruneß is a MSc student in Bioinformatics in the Molecular Bioinformatics group of the Goethe University Frankfurt, Germany. He is interested in data analysis and network modeling.
Contributed image: page 136

Lorenzo Casalino is a post-doctoral scientist in the group of Prof. Rommie Amaro at the University of California San Diego, where he focuses on mesoscale computer simulations and modelling of the Influenza virus. He earned his PhD in Physics and Chemistry of Biological Systems at the International School for Advanced Studies (SISSA, Trieste) under the supervision of Dr. Alessandra Magistrato. His PhD thesis has been focused on RNA splicing and has inspired the artistic rendition reported here.
Contributed image: page 114

Ferran Casbas is an early-stage researcher at the University of Nottingham, with a passion for using novel approaches to solving puzzles and making discoveries. He also likes to program computers to take the strain out of otherwise tedious data analysis.
Contributed image: page 34

George T. Chen is a graduate student at the University of California, Irvine, robotics educator, and graphic designer. He is interested in understanding how cancer cells communicate with each other and their environment.
Contributed image: page 38

Michael Colman is Research Fellow and Junior Lecturer in cardiovascular sciences at the University of Leeds, UK. His research focuses on simulating multi-scale mechanisms of cardiac arrhythmias, with a particular interest in the propagation of microscopic, random fluctuations to the macroscopic scale.
Contributed images: pages 68 and 118

Stephen Coombes is Professor of applied mathematics in the School of Mathematical Sciences at the University of Nottingham. He is interested in the application of principles from nonlinear dynamics and statistical physics to the study of neural systems.
Contributed image: page 32

Diana Ivette Cruz Dávalos is a PhD student at the University of Lausanne. She analyzes the genomes of ancient and modern populations from the Americas to understand their evolution.
Contributed image: page 56

John Dallon is Professor of mathematics at Brigham Young University, Utah. He is interested in cell motion, discrete and stochastic mathematical modeling, and scientific computing.
Contributed image: page 20

Charlotte Deane is Professor at the Department of Statistics at Oxford University, where she is currently Department Head. Her research group focuses on protein informatics, including protein evolution, structure prediction, and interaction networks.
Contributed image: page 82

Andreas Deutsch is Head of the Department of Innovative Methods of Computing at the Centre for Information Services and High Performance Computing at Dresden University of Technology. He is interested in collective behavior in biology and wants to understand how mathematical modeling and analysis can help to understand principles of biological pattern formation.
Contributed image: page 104

Armin Drusko is a PhD student at the Goethe University Frankfurt, working on the analysis of 3D images of moving cells.
Contributed image: page 46

Author Information

Robert A. Edwards is working at the Department of Pathology, Chao Family Comprehensive Cancer Center, Irvine, California.
Contributed image: page 38

Heiko Enderling is associate member in the Departments of Integrated Mathematical Oncology and Radiation Oncology at the Moffitt Cancer Center in Tampa Bay, Florida. He directs a research group on Quantitative Personalized Oncology, which aims to develop mathematical models to predict treatment responses for individual patients with the ultimate goal to optimize precision cancer therapy.
Contributed image: page 130

Nils Faber is a 3D designer and virtual reality researcher in the Delft Starship Team (DSTART) at Delft University of Technology. From making exploratory visualisations, 3D modeling and prototyping, to 3D printing functional models, he loves to make abstract ideas real and tangible. As a visual artist, he explores both the fictional and more practical aspects of human space exploration.
Contributed image: page 134

Laura Fischer studied molecular biotechnology at the University of Heidelberg, Germany. She created this image in the scope of a lab rotation in the group of Franziska Matthäus.
Contributed image: page 96

Stefanos Folias is Associate Professor of mathematics at the University of Alaska Anchorage. His research interests lie in fields of mathematical neuroscience and spatiotemporal dynamical systems.
Contributed image: page 132

Urszula Foryś is Chair of the Section of Biomathematics and Game Theory at the Faculty of Mathematics, Informatics and Mechanics, University of Warsaw, and one of the leading researches in the field of biomathematics in Poland. Biomathematics has been her passion for many years, starting from studies under the supervision of Prof. Wiesław Szlenk.
Contributed image: page 62

Hermann B. Frieboes is Associate Professor in bioengineering at the University of Louisville, focusing on the integration of biological, mathematical, and computational approaches to the study of disease progression and treatment, including cancer.
Contributed image: page 88

Samuel H. Friedman is a Staff Scientist at Opto-Knowledge Systems, Inc. in Torrance, California. He applies mathematics, physics, and computer programming to different fields: galaxies, biological cells, autonomously moving agents, and image processing.
Contributed image: page 42

Stefan Fuertinger is a Scientific Software Developer at Ernst Strüngmann Institute (ESI) for Neuoscience in Cooperation with Max Planck Society. He works in the area of computational biology, mathematical modelling and high-performance computing. He is interested in applying modern mathematical strategies to complex challenges in biomedical applications.
Contributed image: page 58

Jill Gallaher is an applied Research Scientist in the Integrated Mathematical Oncology Department at the Moffitt Cancer Center in Tampa Bay, Florida. She is interested in tumor heterogeneity, drug resistance, and spatial modeling. She enjoys the creativity involved in turning biological systems into mathematical abstractions.
Contributed image: page 76

Chandler Gatenbee is Research Scientist in the Department of Mathematical Oncology at Moffitt Cancer Center in Tampa Bay, Florida. He uses computational models, evolutionary theory, and image analysis to study the eco-evolutionary dynamics of cancer to better understand tumorigenesis, the evolution of resistance, and response to treatment.
Contributed images: pages 70, 124 and 128

Ahmadreza Ghaffarizadeh is a former Postdoctoral Research Associate at the University of Southern California. He is an enthusiast of implementing computational systems that can solve real-world problems.
Contributed image: page 42

Aytül Gökçe was a PhD student at the University of Nottingham in the UK at the time this research was performed. She is now a Research Fellow at Ordu University in Turkey, working in the broad area of applied mathematics. Aytul's research is mainly devoted to dynamical systems modelling, and the understanding of patterning in neural systems.
Contributed image: page 32

Chaitanya Gokhale from the Max Planck Institute for Evolutionary Biology, Plön, Germany, is focused on making use of evolutionary game theory and theoretical ecology to explore sociobiological concepts. Particularly he is interested in the inclusion of non-linearities coming from ecological interactions (antagonistic as well as symbiotic).
Contributed image: page 16

Alexandre Gouy is a PhD student in population genetics at the University of Bern, Switzerland. He is studying how human populations adapted to their environment using both theoretical models and empirical genomic data.
Contributed image: page 56

Sylvain Gretchko is an undergraduate student in mathematics at the University of British Columbia Okanagan, Canada. He is particularly interested in using his software engineering skills to visualize the rich world that emerges from mathematical equations.
Contributed image: page 122

Sara Hamis is studying for a PhD in mathematics at Swansea University, UK. Her research is mathematical, though her motivation stems from an aspiration to advance cancer research.
Contributed image: page 50

Martin-Leo Hansmann is Distinguished Professor of Pathology at the Goethe University, Frankfurt am Main, Germany. His expertise is lymph node pathology including malignant lymphomas. Molecular pathology and laser scanning technologies are applied to understand immunological functions of reactive and neoplastic lymph nodes.
Contributed image: page 28

Sarah Anne Harris is Associate Professor of Biological Physics at the University of Leeds. She has always been interested in understanding how biological systems perform their amazing functions within the confines of the laws of physics.
Contributed as Editor

Randy Heiland is a Research Associate in intelligent systems engineering at Indiana University, Bloomington, USA. He enjoys mentoring students and writing software that helps scientists, engineers, and the lay public gain deeper insights.
Contributed image: page 42

Thomas Hillen is Professor at the University of Alberta, Edmonton, Canada, and he is currently President of the Canadian Applied and Industrial Mathematical Society (CAIMS). He works in mathematical biology and his main motivation is to use mathematical modelling for the common good, such as, for the modelling of cancer and cancer treatments.
Contributed image: page 142

Sandesh Hiremath works as algorithm Engineer for Valeo Schalter und Sensoren. His interests are in studying/modelling/implementing complex systems that behave autonomously, especially biological systems.
Contributed image: page 84

Charlie Hodgman is Emeritus Professor of Bioinformatics and Systems Biology at the School of Biosciences, University of Nottingham, UK. Across both academia and industry, he has built his successful research career on using mathematical and computational techniques to predict biological mechanisms and functions, so that the most incisive experiments can be carried out.
Contributed images: pages 34 and 66

Wim Hordijk is currently a research fellow at the Konrad Lorenz Institute for Evolution and Cognition Research in Klosterneuburg, Austria, and is particularly interested in the origin and evolution of life on earth. He has worked, lived, and traveled all over the world to not only study, but also personally admire the beauty and diversity of nature.
Contributed image: page 44

Thomas Horger is a postdoctoral researcher at the Chair of Numerical Mathematics at the Technical University of Munich. He works on numerics, model order reduction techniques, and mathematical models in engineering, biology and medicine. He is fascinated by the combination of numerics and modeling.
Contributed image: page 24

Nick Jones is Reader in applied mathematics and mathematical physics at the Imperial College of London, UK. His group is interested in mathematical approaches to the natural world, and in particular (bio)energetics, inference and their interface.
Contributed image: page 82

Neha Khetan is a PhD student at the Institute of Science Education and Research (IISER), Pune in India. Emergence of self organized order and form of functional relevance in nature inspires her. She is interested in understanding the interplay of biophysical and selection forces that drive the beautiful and functional patterns in biological systems.
Contributed image: page 86

Author Information

Yong-Jung Kim is Professor at the Korea Advanced Institute of Science and Technology (KAIST). His research interests are in partial differential equation modelling of biological processes, in particular those processes that lead to oriented movement of species. Dr. Kim's research has focussed on directional cues due to orientations towards food sources, avoidance of harmful environment, avoidance of predators, and orientation towards shelter and mates.
Contributed images: pages 52 and 100

Ina Koch is a Professor for molecular bioinformatics at the Institute of Computer Science at the Goethe University, Frankfurt am Main. She is interested in method development and application for computational methods in biology and medicine, in particular using graph-theoretic methods.
Contributed images: pages 28, 112 and 136

Anna Konstorum is a postdoctoral fellow at the Center for Quantitative Medicine, University of Connecticut Health Center, USA.
Contributed image: page 108

Kai Kopfer is a postdoctoral fellow at FIAS, Goethe University Frankfurt. He is interested in cell polarization and develops PDE models coupling chemical and mechanical regulatory processes.
Contributed image: page 126

Petros Koumoutsakos is a Chair for Computational Science at ETH Zurich. He is interested in computing and its application to the interfaces of fluid mechanics, nanotechnology and biology.
Contributed images: pages 18 and 110

Nina Kudryashova has just graduated from the University of Ghent with the thesis devoted to the relation of cardiac tissue patterns to arrhythmias. She enjoys studying in silico how simple rules of nature form complex patterns.
Contributed images: pages 78 and 138

Christina Kuttler is Professor for mathematics in life sciences at the Technical University of Munich, Germany. Since many years, she is working on mathematical modeling of bacterial communication, and still finds it a fascinating research topic.
Contributed image: page 24

Mary Lee is a mathematician working at the Rand Corporation.
Contributed image: page 38

Gaëlle Letort is a computational biologist (postdoc) at the Terret&Verlhac team, CIRB, Collège de France. Computer science and mathematics's huge fan, she tried to apply them to scrutinize cells crazy lives, because of its medical scope but also because it's beautiful.
Contributed images: pages 64, 90, 94 and 140

Anna Lewis completed her PhD at the Systems Biology Doctoral Training Center at the University of Oxford. The art presented here formed part of her thesis. She has since been working in applied genetics as a computational biologist and product manager. She thinks a lot about bioethical questions.
Contributed image: page 82

Jana Lipková is a PhD student at the Technical University of Munich in the Image Based Biomedical Modelling lab. She is interested in medical imaging, computational modelling in medicine and scientific visualisations.
Contributed images: pages 18 and 110

John Lowengrub is Chancellor's Professor of mathematics, biomedical engineering and chemical engineering and materials science at the University of California, Irvine. His research interests include the development of multiscale models of developing biological tissues and tumors.
Contributed images: pages 18, 38 and 108

Jonathan Lynch is Distinguished Professor of plant nutrition at the Pennsylvania State University and a Chair in Root Biology at the University of Nottingham. His research focuses on understanding how we can develop crops with better growth under drought and low soil fertility in order to improve global food security.
Contributed image: page 22

Paul Macklin is an Associate Professor of intelligent systems engineering and Director of Undergraduate Studies at Indiana University. He develops simulation tools to help scientists understand and design biological systems ranging from bacterial colonies to cancer.
Contributed images: pages 42 and 88

Davide Maestrini is a theoretical physicist and postdoctoral researcher in the Mathematical Division at City of Hope National Medical Center, California. He is mainly interested in heavy metal, black metal, and death metal music. In his spare time he is interested in non-linear and statistical mechanics far from the equilibrium applied to biological systems.
Contributed image: page 14

Philip Maini, Fellow of the Royal Society, is Statutory Professor of Mathematical Biology, and Director of the Wolfson Centre for Mathematical Biology, Oxford, UK. His interest is in collective cell motion and in using mathematics to help understand biology. He only ever has been passionate about one thing – football – but that died a few years back.
Contributed image: page 48

Anna Marciniak-Czochra is Professor at the Heidelberg University, Germany. Her work is focused on multi-scale mathematical modeling and analysis of the dynamics of structure formation and self-organization in systems of cells. She is an advocate of exploration of the interface of mathematics and biosciences.
Contributed image: page 30

Jessa Marley is currently an MSc student in applied mathematics at the University of Alberta, Edmonton, Canada, studying shark movement in the Atlantic. She has been drawing throughout her entire life and enjoys finding ways to combine her passions.
Contributed image: page 142

Franziska Matthäus is a research fellow at the Frankfurt Institute of Advanced Sciences (FIAS) in the area of computational biology. She is interested in everything that moves and shares a passion for images arising from scientific research.
Contributed images: pages 26, 46, 96 and 126

Sebastian Matthäus heads a prestigious graphic design agency in Berlin, Germany. He is master of shape, color, light and animation, and turns information into magic beauty. For further information see www.grenzfarben.de.
Contributed as Editor

Bjoern Menze is Assistant Professor of computer Science and medicine at the Technical University of Munich, Germany. His research is in medical image computing, exploring topics at the interface of machine learning, image-based modeling and computational physiology.
Contributed images: pages 18 and 110

Moritz Mercker is a postdoctoral researcher in the area of biological pattern formation, affiliated with the Heidelberg University, Germany. He is interested in tissues and membranes with a focus on mechanochemical interactions. His passion is the question how complex patterns arise from simple structures.
Contributed images: page 30

Roeland M. H. Merks is Professor of mathematical biology at the Mathematical Institute and the Institute for Biology of Leiden University, The Netherlands. Merks is interested in problems of animal and plant development, with a focus on angiogenesis and mechanobiology. He also explores the use of mathematical biology algorithms to produce works of art at www.instagram.com/roelandmerks.
Contributed images: pages 4 and 92

Michael Meyer-Hermann is Head of the Department of Systems Immunology at the Helmholtz Centre for Infection Research in Braunschweig, Germany. He is fascinated by the complexity of living matter and by the capacity of the human brain to reduce this complexity to simple principles.
Contributed image: page 74

Laura Miller is Professor of biology and mathematics at the University of North Carolina at Chapel Hill, in Chapel Hill, NC USA. She has a lab full of squishy invertebrates, which enables her to use an integrative approach to study science through a blend of complementary experiments and computation.
Contributed image: page 106

Shannon M. Mumenthaler is Assistant Professor of medicine at the University of Southern California in LA, USA. She is inspired by working with colleagues from diverse scientific backgrounds to devise new ways of approaching cancer and creating more relevant model systems for drug treatment studies.
Contributed images: pages 42 and 88

François Nédélec is passionate about cellular architecture and studies how filaments are organized inside living cells. He is affiliated with the European Molecular Biology Laboratory (Heidelberg) and the Center for Interdisciplinary Research (Paris).
Contributed image: page 10

Martin A. Nowak is Professor of biology and mathematics at Harvard University and Director of the Program for Evolutionary Dynamics. He works on the mathematical description of evolutionary processes. He is the recipient of numerous prizes and has authored of over 350 papers and four books.
Contributed image: page 2

Aenne Oelker is a postdoctoral researcher at the Chair of Mathematical Modeling at the Technical University of Munich. She is interested in using mathematics to capture the characteristics of complex systems and is fascinated by bacterial pattern formation.
Contributed image: page 24

Author Information

Hans Othmer is a Professor of mathematics, a member of the Digital Technology Center, and an adjunct faculty member in Chemical Engineering at the University of Minnesota. His research interests involve cancer modeling, pattern formation, cell motility and stochastic processes.
Contributed image: page 20

Kevin Painter is a Professor of mathematics at Heriot-Watt University, Edinburgh. He is fascinated by the beautiful patterns that nature seemingly creates with ease, as well as the beautiful mathematics that lie at their basis.
Contributed image: page 40

Amelia Palermo is a post-doctoral scientist at The Scripps Research Institute, San Diego, California, where she investigates the interplay between metabolic networks and biological phenotypes. She develops novel strategies for metabolite imaging and omics data interpretation for systems biology. She completed her doctoral studies at the University of Rome "La Sapienza" and at the Swiss Federal Institute of Technology ETH Zurich.
Contributed image: page 114

Giulia Palermo is an Assistant Professor at the University of California Riverside (https://palermolab.com). Her research interfaces molecular dynamics simulations and cryo-electro-microscopy refinement techniques, to study increasingly realistic biological systems involved in genome editing and regulation. Her work has been featured in the 2018 J. Am. Chemistry. Soc. Young Investigators Virtual Issue as one of the most outstanding work from young investigators.
Contributed image: page 114

Alexander Panfilov is the leader of the Cardiac Biophysics Group at Ghent University. He is interested in spatial pattern dynamics in biology and medicine related to the heart and development.
Contributed images: pages 78 and 138

Nick Parayko is an M.Sc. Student in Ecology at the University of Alberta, studying responses of Ferruginous Hawks to anthropogenic change. He is a keen birder and has a passion for ornithology and wildlife photography.
Contributed image: page 142

Monika J. Piotrowska is a vice-Director of the Institute of Applied Mathematics and Mechanics at the Faculty of Mathematics, Informatics and Mechanics, University of Warsaw and a former master student of prof. Urszula Foryś. Her main scientific interests focus on modeling of biomedical phenomena using both continuous models and computational approaches.
Contributed image: page 62

Jan Poleszczuk is Assistant Professor and the Head of the Laboratory of Mathematical Modelling of Biomedical Systems at the Nalecz Institute of Biocybernetics and Biomedical Engineering in Poland. He is constantly looking for medical and biological problems that can be tackled using mathematical modeling, frequently asking a lot of questions in the process.
Contributed images: pages 62 and 130

Mason A. Porter is Professor in the Department of Mathematics at the University of California Los Angeles. His work covers a broad spectrum of applied mathematics, with a focus on complex systems, networks, and nonlinear systems. His research interests were sparked initially by what he found visually appealing.
Contributed image: page 82

Gibin Powathil is Associate Professor at the Department of Mathematics, Swansea University, UK. He is an applied mathematician focusing on interdisciplinary multi-scale approaches to understand the underlying complexity of cancer.
Contributed image: page 50

Eric Puttock is working at the Department of Mathematics, University of California, Irvine, California.
Contributed image: page 38

Jagath C. Rajapakse is Professor of computer engineering at the Nanyang Technological University, Singapore. His research interests are in machine and deep learning and their applications to biology and medicine. He has passions for unwinding molecular networks and understanding the brain through neuroimages.
Contributed image: page 72

Elisabeth Rens is currently a postdoc in the mathematical biology group of the University of British Columbia in Vancouver, Canada. She is interested in multiscale modeling of cell-cell interactions in development and has a passion for making models come alive with colourful graphics.
Contributed image: page 92

Jens Rieser is a PhD student in the Molecular Bioinformatics group of the Goethe University Frankfurt, Germany. He is interested in network modeling ranging from visualization to analysis.

Contributed images: pages 112 and 136

Mark Robertson-Tessi is a research scientist in the Integrated Mathematical Oncology department at Moffitt Cancer Center, Tampa Bay, USA. He enjoys blending artistic and scientific approaches to problem solving.

Contributed image: page 60

Russell Rockne is a Mathematician, Artist, and Musician. He is interested in everything curious and beautiful, from equations to extracellular galaxies. He dedicates this work to his wife Sarah and his children Chloe, Ella, and Alex.

Contributed images: pages 14 and 98

Diego Rossinelli is Chief Technical Officer at Lucid Concepts AG, Zurich, Switzerland, he does Research and Development on image analysis and computational science at the large scale.

Contributed images: pages 18 and 110

Prativa Sahoo is a postdoc at the City of Hope National Medical Centre, Duarte, California, in the field of mathematical oncology. She is interested in only two things: mathematics and art. She likes to visualize art in mathematics and mathematics in art.

Contributed image: page 14

Hendrik Schäfer is a PhD student in the Molecular Bioinformatics Group at the Goethe University, Frankfurt. He participates in computer-aided methods to analyse, understand, and visualize medical data.

Contributed image: page 28

Ryan Schenck is a D.Phil Student in genomic medicine and statistics at the Wellcome Trust Centre for Human Genetics, University of Oxford. He is passionate about exploring evolutionary dynamics in somatic tissue and how these dynamics may lead to the development of cancer. Through the merging of mechanistic mathematical modeling, bioimage informatics, and bioinformatics, Ryan is currently undertaking a multifaceted approach to understand how mutations and eventual cell fates interplay within the microenvironment and give rise to field cancerization and neoplasia.

Contributed image: page 116

Hanna Schenk did her PhD supervised by Arne Traulsen in the Department of Evolutionary Theory (Max-Planck-Institute for Evolutionary Biology, Plön, Germany). She used mathematical models and simulations to understand host-parasite co-evolution, focusing on the influence of stochasticity, ecology and various mathematical assumptions.

Contributed image: page 16

Santiago Schnell is the John A. Jacquez Collegiate Professor of Physiology at the University of Michigan Medical School. He is also Professor of Computational Medicine & Bioinformatics at the same institution. Dr. Schnell develops novel approaches to measure reactions and other processes in the biomedical sciences. Research in measurement science facilitates the translation of basic science and clinical research to inspire breakthroughs in the biomedical sciences. He is passionate about integrating Art into Science, Technology, Engineering and Math - the STEM subjects – to transform STEM into STEAM.

Contributed image: page 54

Linus Schumacher is a research group leader in computational biology at the Centre for Regenerative Medicine in Edinburgh, UK. He studies the collective behaviour of molecules, cells, and organisms, with a particular focus on cell populations in development and regeneration.

Contributed image: page 48

Kristina Simonyan is Director of Laryngology Research and Head of the NIH-funded Dystonia and Speech Motor Control Laboratory at Massachusetts Eye and Ear, Associate Neuroscientist at Massachusetts General Hospital and Associate Professor of Otolaryngology Head and Neck Surgery at Harvard Medical School, Boston, USA. She is dedicated to understand the neural mechanisms of normal and diseased speech motor control and other complex voluntary motor behaviors.

Contributed image: page 58

Stefanie Sonner is Assistant Professor in applied analysis at Radboud University Nijmegen, The Netherlands. Her main research interests are in the field of infinite dimensional dynamical systems and nonlinear parabolic systems arising in mathematical biology.

Contributed image: page 84

Jessica L. Sparks is Associate Professor of chemical, paper, and biomedical engineering at Miami University in Oxford, Ohio, USA. Her research interests include soft tissue biomechanics, tissue engineering, and 3D printing of soft materials.

Contributed image: page 88

Author Information

Dov Stekel is Professor of computational biology at the University of Nottingham, UK. His research focuses on using mathematical, computing and statistical techniques to build predictive models for biological systems, with the aims to deepen our understanding of biological systems and to inform the design of future experimental work.
Contributed image: page 66

Damian Stichel is a research fellow at the Deutsche Krebsforschungs Zentrum (DKFZ) in Heidelberg, Germany, in the area of mathematics and bioinformatics. His interest focuses on applying mathematics to solve biological questions.
Contributed image: page 26

Maximilian Strobl is a PhD student at the University of Oxford, UK, studying the drug resistance in ovarian cancer. He is jointly supervised by Professor Philip Maini at Oxford, and Dr Alexander Anderson and Dr Jose Conejo-Garcia at the Moffitt Cancer Center, USA. Maxi is very interested in combining data analysis and mathematical modelling to improve our understanding of the eco-systems in our body, in health and disease.
Contributed images: pages 6, 124 and 128

Valerii M. Sukhorukov was working at the Helmholtz Centre for Infection Research, Braunschweig, Germany. He is interested in examining the key properties of living cells by reproducing them with computational algorithms on the border between nonlinear systems and emergent technologies.
Contributed image: page 74

Christina Surulescu is Professor for mathematics with applications to biology and medicine at the Technical University of Kaiserslautern, Germany. She and her group are particularly interested in partial differential equation models for motion of cells in interaction with their surroundings, with an emphasis on modeling tumor development, invasion, and therapy approaches.
Contributed image: page 84

Arne Traulsen is working at the Max-Planck-Institute for Evolutionary Biology, Plön, Germany, on mathematical and computational models that aim to improve our understanding of evolutionary processes. The applications of these models considered in the department include coevolutionary processes, the evolution of cooperation, host-microbiome interactions and the somatic evolution of cancer.
Contributed image: page 16

Angelo C.J. Vermeulen is a biologist and space systems researcher at Delft University of Technology. He is the founder of DSTART (Delft Starship Team) and develops bio-inspired concepts for interstellar exploration. In collaboration with the European Space Agency, he works on closed-loop ecosystems for long-duration space missions. Vermeulen is also a visual artist and co-creates installations that blend art, science and technology.
Contributed image: page 134

Bartlomiej Waclaw is a Royal Society of Edinburgh Research Fellow and a Reader at the University of Edinburgh. He is a biological physicist interested in Darwinian evolution of bacteria and cancer, and has a passion for computer graphics.
Contributed image: page 2

Kehui Wang is working at the Department of Pathology, Chao Family Comprehensive Cancer Center, Irvine, California.
Contributed image: page 38

Marian L. Waterman works at the Department of Microbiology and Molecular Genetics, Cancer Research Institute / Chao Family Comprehensive Cancer Center, Center for Complex Biological Systems, Cancer Systems Biology @ UC Irvine.
Contributed image: page 38

Glenn Webb is a Professor of mathematics at Vanderbilt University in Nashville, Tennessee, USA. His research concerns population models in biology and medicine. He views the world in terms of populations, with complex relationships and interactions, all amenable to mathematical models.
Contributed images: pages 12, 36, 80 and 102

Huaming Yan is a postdoctoral researcher at University of California, Irvine. He is interested in using mathematical models to study cancers and find novel cancer therapies.
Contributed images: pages 38 and 108

Anna Zhigun is Lecturer in Applied Mathematics at the Queen's University Belfast, UK, working in the area of analysis for PDE systems arising in biology. She has an eye for constructing solutions to new and intricate models and understanding patterns they produce.
Contributed image: page 84